Galois Theory, Coverings, and Riemann Surfaces

Askold Khovanskii

Galois Theory, Coverings, and Riemann Surfaces

 Springer

Askold Khovanskii
Dept. Mathematics
University of Toronto
Toronto, Ontario, Canada

ISBN 978-3-662-51956-1   ISBN 978-3-642-38841-5 (eBook)
DOI 10.1007/978-3-642-38841-5
Springer Heidelberg New York Dordrecht London

Mathematics Subject Classification (2010): 55-02, 12F10, 30F10

© Springer-Verlag Berlin Heidelberg 2013
Softcover reprint of the hardcover 1st edition 2013
Translation of Russian edition entitled "Teoriya Galua, Nakrytiya i Rimanovy Poverkhnosti", published
by MCCME, Moscow, Russia, 2006

Printed on acid-free paper

Springer is part of Springer Science+Business Media (www.springer.com)

Preface

The main goal of this book is an exposition of Galois theory and its applications to the questions of solvability of algebraic equations in explicit form. Apart from the classical problem on solvability of an algebraic equation by radicals, we also consider other problems of this type, for instance, the question of solvability of an equation by radicals and by solving auxiliary equations of degree at most k.

There exists a surprising analogy between the fundamental theorem of Galois theory and classification of coverings over a topological space. A description of this analogy is the second goal of the present book. We consider several classifications of coverings closely related to each other. At the same time, we stress a formal analogy between the results thus obtained and the fundamental theorem of Galois theory. Apart from coverings, we consider finite ramified coverings over Riemann surfaces (i.e., over one-dimensional complex manifolds). Ramified coverings are slightly more complicated than unramified finite coverings, but both types of coverings are classified in the same way.

The third goal of the book is a geometric description of finite algebraic extensions of the field of meromorphic functions on a Riemann surface. For such surfaces, the geometry of ramified coverings and Galois theory are not only analogous but in fact very closely related to each other. This relationship is useful in both directions. On the one hand, Galois theory and Riemann's existence theorem allow one to describe the field of functions on a ramified covering over a Riemann surface as a finite algebraic extension of the field of meromorphic functions on the Riemann surface. On the other hand, the geometry of ramified coverings together with Riemann's existence theorem allows one to give a transparent description of algebraic extensions of the field of meromorphic functions over a Riemann surface.

The book is organized as follows. The first chapter is devoted to Galois theory. It is absolutely independent of the other chapters. It can be read separately.

The second chapter is devoted to coverings over topological spaces and to ramified coverings over Riemann surfaces. It is almost independent of the first chapter. In the second chapter, we stress a formal analogy between the classification of coverings and the fundamental theorem of Galois theory. This is the only connection between the chapters—to read the second chapter, it is enough to know the formulation of the fundamental theorem of Galois theory.

The third chapter relies on Galois theory as well as on the classification of ramified coverings over Riemann surfaces. Nevertheless, it can also be read independently if the reader accepts without proof the necessary results of the first two chapters.

The numbering of theorems, propositions, lemmas etc. in every chapter is separate, but formulas are numbered consistently through the whole book.

The book is addressed to mathematicians and to undergraduate and graduate students majoring in mathematics. Some results of the book (for instance, necessary conditions of various forms of solvability of complicated algebraic equations via solutions of simpler algebraic equations, a description of an analogy between the theory of coverings and Galois theory) might be of interest to experts.

Contents

Chapter 1
Galois Theory

In this first chapter, we give an exposition of Galois theory and its applications to the questions of solvability of algebraic equations in explicit form.

In Sects. 1.1–1.4, we consider a field P on which a finite group G acts by field automorphisms. Elements of the field P fixed under the action of G form a subfield $K \subseteq P$, which is called the invariant subfield.

In Sect. 1.1, we show that if the group G is solvable, then the elements of the field P are representable by radicals through the elements of the invariant field K. (Here, an additional assumption is needed that the field K contains all roots of unity of order equal to the cardinality of G.) If P is the field of rational functions of n variables, then G is the symmetric group acting by permutations of the variables, and K is the subfield of symmetric functions of n variables. This result provides an explanation for the fact that algebraic equations of degrees 2 to 4 in one variable are solvable by radicals.

In Sect. 1.2, we show that for every subgroup G_0 of the group G, there exists an element $x \in P$ whose stabilizer is equal to G_0. The results of Sects. 1.1 and 1.2 are based on simple considerations from group theory; they use an explicit formula for the Lagrange interpolating polynomial.

In Sect. 1.3, we show that every element of the field P is algebraic over the field K. We prove that if the stabilizer of a point $z \in P$ contains the stabilizer of a point $y \in P$, then z is the value at y of some polynomial over the field K. This proof is also based on the study of the Lagrange interpolating polynomial (see Sect. 1.3.3).

In Sect. 1.4, we introduce the class of k-solvable groups. We show that if a group G is k-solvable, then the elements of the field P are representable in k-radicals (i.e., can be obtained by taking radicals and solving auxiliary algebraic equations of degree k or less) through the elements of the field K. (Here, we also need to assume additionally that the field K contains all roots of unity of order equal to the cardinality of G.)

Consider now a different situation. Suppose that a field P is obtained from a field K by adjoining all roots of a polynomial equation over K with no multiple roots. In this case, there exists a finite group G of automorphisms of the field P whose invariant field coincides with K. To construct the group G, the initial equa-

A. Khovanskii, *Galois Theory, Coverings, and Riemann Surfaces*,
DOI 10.1007/978-3-642-38841-5_1, © Springer-Verlag Berlin Heidelberg 2013

tion needs to be replaced with an equivalent Galois equation, i.e., with an equation each of whose roots can be expressed through any other root (see Sect. 1.5). This group G of automorphisms is constructed in Sect. 1.6.

Thus Sects. 1.2, 1.3, 1.5, and 1.6 contain proofs of the central theorems of Galois theory. In Sect. 1.7, we summarize, then state and prove, the Fundamental theorem of Galois theory.

An algebraic equation over a field is solvable by radicals if and only if its Galois group is solvable (Sect. 1.8), and it is solvable by k-radicals if and only if its Galois group is k-solvable (Sect. 1.9). In Sect. 1.10, we discuss the question of solvability of algebraic equations with higher complexity by solving equations with lower complexity. We give a necessary condition for such solvability in terms of the Galois group of the equation.

In Sect. 1.11, we classify finite fields. We check that the fundamental theorem of Galois theory holds for finite fields (the proof of the fundamental theorem given in Sect. 1.7 does not go through for finite fields).

In this chapter, a major focus is on the applications of Galois theory to problems of solvability of algebraic equations in explicit form. However, the exposition of Galois theory does not refer to these applications. The fundamental principles of Galois theory are covered in Sects. 1.2, 1.3, 1.5–1.7. These sections can be read independently of the rest of the chapter.

A recipe for solving algebraic equations by radicals (including solutions of general equations of degree 2 to 4) is given in Sect. 1.1, and is independent of the rest of the text. The classification of finite fields given in Sect. 1.11 is also practically independent of the rest of the text. These sections can be read independently of the rest of the chapter.

1.1 Action of a Solvable Group and Representability by Radicals

In this section, we prove that if a finite solvable group G acts on a field P by field automorphisms, then (under certain additional assumptions on the field P), all elements of P can be expressed through the elements of the invariant field K by radicals and arithmetic operations.

A construction of a representation by radicals is based on linear algebra (see Sect. 1.1.1). In Sect. 1.1.2, we use this result to prove solvability of equations of low degrees. To obtain explicit solutions, the linear algebra construction needs to be done explicitly. In Sect. 1.1.3, we introduce the technique of Lagrange resolvents, which allows us to perform an explicit diagonalization of an Abelian linear group. In Sect. 1.1.4, we explain, how Lagrange resolvents can help to write down explicit formulas with radicals for the solutions of equations of degree 2 to 4.

The results of this section are applicable in the general situation considered in Galois theory. If a field P is obtained from the field K by adjoining all roots of an algebraic equation without multiple roots, then there exists a group G of automorphisms of the field P whose invariant field is the field K (see Sect. 1.7.1). This group is called the *Galois group of the equation*. It follows from the results of this section that an equation whose Galois group is solvable can be solved by radicals (the sufficient condition for solvability by radicals from Theorem 1.8.7). The existence of the Galois group is by no means obvious; it is one of the central results of Galois theory. In this section, we do not prove this theorem (a proof is given in Sect. 1.7.1); we assume from the very beginning that the group G exists.

In a variety of important cases, the group G is given a priori. This is the case, for example, if K is the field of rational functions of a single complex variable, P is the field obtained by adjoining to K all solutions of an algebraic equation, and G is the monodromy group of the algebraic function defined by this equation (see Chap. 3).

1.1.1 A Sufficient Condition for Solvability by Radicals

The fact that we deal with fields is barely used in the construction of a representation by radicals. To emphasize this, we describe this construction in a general setup, whereby a field is replaced with an algebra V, which may even be noncommutative. (In fact, we do not even need to multiply different elements of the algebra. We will use only the operation of taking an integer power k of an element and the fact that this operation is homogeneous of degree k under multiplication by elements of the base field: $(\lambda a)^k = \lambda^k a^k$ for all $a \in V$, $\lambda \in K$.)

Let V be an algebra over a field K containing all nth roots of unity for some integer $n > 0$. A finite Abelian group of linear transformations of a finite-dimensional vector space over the field K can be diagonalized in a suitable basis (see Sect. 1.1.3) if the order of the group is not divisible by the field characteristic (for fields of zero characteristic, no restrictions on the order of the group are necessary). In particular, we have the following proposition.

Proposition 1.1.1 *Let G be a finite Abelian group of order n acting by automorphisms of the algebra V. Assume that the order n is not divisible by the field characteristic. Suppose that K contains all nth roots of unity. Then every element of the algebra V is representable as a sum of $k \le n$ elements $x_i \in V$, $i = 1, \ldots, k$, such that x_i^n lies in the invariant subalgebra V_0.*[1]

Proof Consider a finite-dimensional vector subspace L in the algebra V spanned by the G-orbit of an element x. The space L splits into a direct sum $L = L_1 \oplus \cdots \oplus L_k$ of eigenspaces for all operators from G (see Sect. 1.1.3). Therefore, the vector x can be represented in the form $x = x_1 + \cdots + x_k$, where $x_1, \ldots, x_k$ are eigenvectors for all operators from the group. The corresponding eigenvalues are degree-n roots of unity. Therefore, the elements $x_1^n, \ldots, x_k^n$ belong to the invariant algebra V_0. $\square$

---

[1] That is, the fixed-point set of the group G (*translator's note*).

Definition 1.1.2 We say that an element x of the algebra V is an *nth root of an element a* if $x^n = a$.

We can now restate Proposition 1.1.1 as follows: every element x of the algebra V is representable as a sum of nth roots of some elements of the invariant subalgebra.

Theorem 1.1.3 *Let G be a finite solvable group of automorphisms of the algebra V of order n. Suppose that the field K contains all the nth roots of unity. Then every element x of the algebra V can be obtained from the elements of the invariant subalgebra V_0 by root extractions and summations.*

We first prove the following simple statement about an action of a group on a set. Suppose that a group G acts on a set X, that H is a normal subgroup of G, and that X_0 is the subset of X consisting of all points fixed under the action of G.

Proposition 1.1.4 *The subset X_H of the set X consisting of the fixed points under the action of the normal subgroup H is invariant under the action of G. There is a natural action of the quotient group G/H with the fixed-point set X_0.*

Proof Suppose $g \in G$, $h \in H$. Then the element $g^{-1}hg$ belongs to the normal subgroup H. Let $x \in X_H$. Then $g^{-1}hg(x) = x$, or $h(gx) = g(x)$, which means that the element $g(x) \in X$ is fixed under the action of the normal subgroup H. Thus the set X_H is invariant under the action of the group G. Under the action of G on X_H, all elements of H correspond to the identity transformation. Hence the action of G on X_H reduces to an action of the quotient group G/H. $\square$

We now proceed with the proof of Theorem 1.1.3.

Proof Since the group G is solvable, it has a chain of nested subgroups $G = G_0 \supset \cdots \supset G_m = e$, in which the group G_m consists of the identity element e only, and every group G_i is a normal subgroup of the group G_{i-1}. Moreover, the quotient group G_{i-1}/G_i is Abelian.

Denote by $V_0 \subset \cdots \subset V_m = V$ the chain of invariant subalgebras of the algebra V with respect to the action of the groups $G_0, \ldots, G_m$. By Proposition 1.1.4, the Abelian group G_{i-1}/G_i acts naturally on the invariant subalgebra V_i, leaving the subalgebra V_{i-1} pointwise fixed. The order m_i of the quotient group G_{i-1}/G_i divides the order of the group G. Therefore, Proposition 1.1.1 is applicable to this action. We conclude that every element of the algebra V_i can be expressed with the help of summation and root extraction through the elements of the algebra V_{i-1}. Repeating the same argument, we will be able to express every element of the algebra V through the elements of the algebra V_0 by a chain of summations and root extractions. $\square$

1.1.2   The Permutation Group of the Variables and Equations of Degree 2 to 4

Theorem 1.1.3 explains why equations of low degree are solvable by radicals.

Suppose that the algebra V is the polynomial ring in the variables $x_1, \ldots, x_n$ over the field K. The symmetric group $S(n)$ consisting of permutations of n elements acts on this ring, permuting the variables $x_1, \ldots, x_n$ in polynomials from this ring. The invariant algebra of this action consists of all symmetric polynomials. Every symmetric polynomial can be represented explicitly as a polynomial of the elementary symmetric functions $\sigma_1, \ldots, \sigma_n$, where $\sigma_1 = x_1 + \cdots + x_n$, $\sigma_2 = \sum_{i<j} x_i x_j, \ldots, \sigma_n = x_1 \cdots x_n$.

Consider the general algebraic equation $x^n + a_1 x^{n-1} + \cdots + a_n = 0$ of degree n. According to Viète's formulas, the coefficients of this equation are equal up to a sign to the elementary symmetric functions of its roots $x_1, \ldots, x_n$. Namely, $\sigma_1 = -a_1, \ldots, \sigma_n = (-1)^n a_n$.

For $n = 2, 3, 4$, the group $S(n)$ is solvable. Suppose that the field K contains all roots of unity of degree 4 and less. Applying Theorem 1.1.3, we obtain that every polynomial of $x_1, \ldots, x_n$ can be expressed through the elementary symmetric polynomials $\sigma_1, \ldots, \sigma_n$ using root extraction, summation, and multiplication by rational numbers. Therefore, Theorem 1.1.3 *for $n = 2, 3, 4$ proves the representability of the roots of a degree-n algebraic equation through the coefficients of this equation using root extractions, summation, and multiplication by rational numbers.*

To obtain explicit formulas for the roots, we need to repeat all the arguments, performing all necessary constructions explicitly. We will do this in Sects. 1.1.3 and 1.1.4.

1.1.3   Lagrange Polynomials and Commutative Matrix Groups

Let T be a polynomial of degree n with leading coefficient 1 over an arbitrary field K. Suppose that the polynomial T has exactly n distinct roots $\lambda_1, \ldots, \lambda_n$. With every root λ_i, we associate the polynomial

$$T_i(t) = \frac{T(t)}{T'(\lambda_i)(t - \lambda_i)}.$$

This polynomial T_i is the unique polynomial of degree at most $n - 1$ that is equal to 1 at the root λ_i and to zero at all other roots of the polynomial T. Let $c_1, \ldots, c_n$ be any collection of elements of the field K. The polynomial $L(t) = \sum c_i T_i(t)$ is called the Lagrange interpolating polynomial with interpolation points $\lambda_1, \ldots, \lambda_n$ and initial values $c_1, \ldots, c_n$. This is the unique polynomial of degree at most n that takes the value c_i at every point λ_i, $i = 1, \ldots, n$.

Consider a vector space V (possibly infinite-dimensional) over the field K and a linear operator $A : V \to V$. Suppose that the operator A satisfies a linear equation

$T(A) = A^n + a_1 A^{n-1} + \cdots + a_{n-1} A + a_n E = 0$, where $a_i \in K$, and E is the identity operator. Assume that the polynomial $T(t) = t^n + a_1 t^{n-1} + \cdots + a_n$ has n distinct roots $\lambda_1, \ldots, \lambda_n$ in the field K. The operator $L_i = T_i(A)$, where

$$T_i(t) = \frac{T(t)}{T'(\lambda_i)(t - \lambda_i)},$$

will be called the *generalized Lagrange resolvent* of the operator A corresponding to the root λ_i. For every vector $x \in V$, the vector $x_i = L_i x$ will be called the generalized Lagrange resolvent (corresponding to the root λ_i) of the vector x.

Proposition 1.1.5

1. *Generalized Lagrange resolvents L_i of the operator A satisfy the following relations*: $L_1 + \cdots + L_n = E$, $L_i L_j = 0$ *for* $i \neq j$, $L_i^2 = L_i$, $A L_i = \lambda_i L_i$.
2. *Every vector $x \in V$ is representable as the sum of its generalized Lagrange resolvents, i.e., $x = x_1 + \cdots + x_n$. Moreover, the nonzero resolvents x_i of the vector x are linearly independent and are equal to eigenvectors of the operator A with the corresponding eigenvalues λ_i.*

Proof

1. Let $\Lambda = \{\lambda_i\}$ be the set of all roots of the polynomial T. By definition, the polynomial T_i is equal to 1 at the point λ_i and is equal to zero at all other points of this set. It is obvious that the following polynomials vanish on the set Λ: $T_1 + \cdots + T_n - 1$, $T_i T_j$ for $i \neq j$, $T_i^2 - T_i$, $t T_i - \lambda_i T_i$. Therefore, each of the polynomials mentioned above is divisible by the polynomial T, which has simple roots at the points of the set Λ. Since the polynomial T annihilates the operator A, i.e., $T(A) = 0$, this implies the relations $L_1 + \cdots + L_n = E$, $L_i L_j = 0$ for $i \neq j$, $L_i^2 = L_i$, $A L_i = \lambda_i L_i$.
2. The second part of the statement is a formal consequence of the first. Indeed, since $E = L_1 + \cdots + L_n$, every vector x satisfies $x = L_1 x + \cdots + L_n x = x_1 + \cdots + x_n$. Assume that the vector x is nonzero, and that some linear combination $\sum \mu_j x_j$ of the vectors $x_1, \ldots, x_n$ vanishes. Then $0 = L_i \sum \mu_j L_j x = \sum L_i L_j \mu_j x = \mu_i x_i$, i.e., every nonzero vector x_i enters this linear combination with the zero coefficient $\mu_i = 0$. The identity $A L_i = \lambda_i L_i$ implies that $A L_i x = \lambda_i L_i x$, i.e., either the vector $x_i = L_i x$ is an eigenvector of L_i with the eigenvalue λ_i, or $x_i = 0$. $\qquad\square$

The explicit construction for the decomposition of x into eigenvectors of the operator A carries over automatically to the case of several commuting operators. Let us discuss the case of two commuting operators in more detail. Suppose that along with the linear operator A on the space V, we are given another linear operator $B : V \to V$ that commutes with A and satisfies a polynomial relation of the form $Q(B) = B^k + b_1 B^{k-1} + \cdots + b_k E = 0$, where $b_i \in K$. Assume that the polynomial $Q(t) = t^k + b_1 t^{k-1} + \cdots + b_k$ has k distinct roots $\mu_1, \ldots, \mu_k$ in the field K. With a root μ_j, we associate the polynomial $Q_j(t) = Q(t)/Q'(\mu_j)(t - \mu_j)$ and the operator $Q_j(B)$, i.e., the generalized Lagrange resolvent of the operator B corresponding to the root μ_j. We call the operator $L_{i,j} = T_i(A)Q_j(B)$ the generalized Lagrange resolvent of the operators A and B corresponding to the pair of roots λ_i, μ_j. The vector $x_{i,j} = L_{i,j}x$ will be called the generalized Lagrange resolvent of the vector $x \in V$ (corresponding to the pair of roots λ_i and μ_j) with respect to the operators A and B.

Proposition 1.1.6

1. *Generalized Lagrange resolvents $L_{i,j}$ of commuting operators A and B satisfy the following relations:* $\sum L_{i,j} = E$, $L_{i_1,j_1}L_{i_2,j_2} = 0$ *for* $(i_1, j_1) \neq (i_2, j_2)$, $L_{i,j}^2 = L_{i,j}$, $AL_{i,j} = \lambda_i L_{i,j}$, $BL_{i,j} = \mu_j L_{i,j}$.
2. *Every vector $x \in V$ is representable as the sum of its generalized Lagrange resolvents, i.e., $x = \sum x_{i,j}$. Moreover, nonzero resolvents $x_{i,j}$ of the vector x are linearly independent, and are equal to eigenvectors of the operators A and B with the eigenvalues λ_i and μ_j, respectively.*

Proof To prove the first part of the proposition, it suffices to multiply the corresponding identities for the generalized resolvents of the operators A and B. The second part of the proposition is a formal consequence of the first part. □

We can now apply the propositions just proved to an operator A of finite order: $A^n = E$. Generalized Lagrange resolvents for such operators are particularly important for solving equations by radicals. These are the resolvents that Lagrange discovered, and we call them the *Lagrange resolvents* (omitting the word "generalized"). Suppose that the field K contains n roots of unity $\xi_1, \ldots, \xi_n$ of degree n, $\xi^n = 1$. By our assumption, $T(A) = 0$, where $T(t) = t^n - 1$. Let us now compute the Lagrange resolvent corresponding to the root $\xi_i = \xi$. We have

$$T_i(t) = \frac{t^n - \xi^n}{n\xi^{n-1}(t - \xi)} = \frac{1}{n\xi^{n-1}}\left(t^{n-1} + \cdots + \xi^{n-1}\right) = \frac{1}{n}\left((\xi^{-1}t)^{n-1} + \cdots + 1\right).$$

The Lagrange resolvent $T_i(A)$ of the operator A corresponding to a root $\xi_i = \xi$ will be denoted by $R_\xi(A)$. We obtain $R_\xi(A) = \frac{1}{n}\sum_{0 \leq k < n}\xi^{-k}A^k$.

Corollary 1.1.7 *Consider a vector space V over a field K containing all the nth roots of unity. Suppose that an operator A satisfies the relation $A^n = E$. Then, for every vector $x \in V$, either the Lagrange resolvent $R_\xi(A)(x)$ is zero, or it is equal to an eigenvector of the operator A with the eigenvalue ξ. The vector x is the sum of all its Lagrange resolvents.*

Remark 1.1.8 Corollary 1.1.7 can be verified directly, without any reference to preceding results.

Let G be a finite group of linear operators on a vector space V over the field K. Let n denote the order of the group G. Suppose that the field K contains all nth roots of unity for some n. Then the space V is a direct sum of subspaces that are eigenspaces simultaneously for all operators from the group G. Let us make this statement more precise. Suppose that the group G is the direct sum of k cyclic groups of orders $m_1, \ldots, m_k$. Suppose that the operators $A_i \in G, \ldots, A_k \in G$ generate these cyclic subgroups. In particular, $A_1^{m_1} = E, \ldots, A_k^{m_k} = E$. For every collection $\lambda = \lambda_1, \ldots, \lambda_k$ of roots of unity of degrees $m_1, \ldots, m_k$, consider the *joint Lagrange resolvent* $L_\lambda = L_{\lambda_1}(A_1) \cdots L_{\lambda_k}(A_k)$ of all generators $A_1, \ldots, A_k$ of the group G.

Corollary 1.1.9 *Every vector $x \in V$ is representable in the form $x = \sum L_\lambda x$. Each of the vectors $L_\lambda x$ is either zero or a common eigenvector of the operators $A_1, \ldots, A_k$ with the respective eigenvalues $\lambda_1, \ldots, \lambda_k$.*

1.1.4 Solving Equations of Degree 2 to 4 by Radicals

In this subsection, we revisit equations of low degree (see Sect. 1.1.2). We will use the technique of Lagrange resolvents and explain how the solution scheme for equations from Sect. 1.1.2 can be used to produce explicit formulas. The formulas themselves will not be written down. We use notation from Sects. 1.1.2 and 1.1.3. Lagrange resolvents of operators will be labeled by the eigenvalues of these operators. Joint Lagrange resolvents of pairs of operators will be labeled by pairs of the corresponding eigenvalues.

Equations of the Second Degree   Assume that the characteristic of a field K is not equal to 2. The polynomial ring $K[x_1, x_2]$ carries a linear action of the permutation group $S(2) = \mathbb{Z}_2$ of two elements. This group consists of the identity map and some operator of order 2. The element x_1 has two Lagrange resolvents with respect to the action of this operator:

$$R_1 = \frac{1}{2}(x_1 + x_2) = \frac{1}{2}\sigma_1 \quad \text{and} \quad R_{-1} = \frac{1}{2}(x_1 - x_2).$$

The square of the Lagrange resolvent R_{-1} is a symmetric polynomial. We have

$$R_{-1}^2 = \frac{1}{4}\big((x_1 + x_2)^2 - 4x_1 x_2\big) = \frac{1}{4}(\sigma_1^2 - 4\sigma_2).$$

We obtain a representation of the polynomial x_1 through the elementary symmetric polynomials

$$x_1 = R_1 + R_{-1} = \frac{\sigma_1 \pm \sqrt{\sigma_1^2 - 4\sigma_2}}{2},$$

which gives the usual formula for the solutions of a quadratic equation.

Equations of the Third Degree   Assume that the characteristic of a field K is not equal to 2 and that K contains the three cube roots of unity. On the polynomial ring $K[x_1, x_2, x_3] = V$, there is an action of the permutation group $S(3)$. The alternating group $A(3)$, which is a cyclic group of order 3, is a normal subgroup of the group $S(3)$. The group $A(3)$ is generated by the operator B defining the permutation x_2, x_3, x_1 of the variables x_1, x_2, x_3. The quotient group $S(3)/A(3)$ is a cyclic group of order 2. Denote by V_1 the invariant subalgebra of the group $A(3)$ (consisting of all polynomials that remain unchanged under all even permutations of the variables) and by V_2 the algebra of symmetric polynomials. The element x_1 has three Lagrange resolvents with respect to the generator B of the group $A(3)$:

$$R_1 = \frac{1}{3}(x_1 + x_2 + x_3),$$

$$R_{\xi_1} = \frac{1}{3}\left(x_1 + \xi_2 x_2 + \xi_2^2 x_3\right),$$

$$R_{\xi_2} = \frac{1}{3}\left(x_1 + \xi_1 x_2 + \xi_1^2 x_3\right),$$

where $\xi_1, \xi_2 = \frac{1}{2}(-1 \pm \sqrt{-3})$ are the cube roots of unity different from one.

We have $x_1 = R_1 + R_{\xi_1} + R_{\xi_2}$, and $R_1^3, R_{\xi_1}^3, R_{\xi_2}^3$ lie in the algebra V_1. Moreover, the resolvent R_1 is a symmetric polynomial, and the polynomials $R_{\xi_1}^3$ and $R_{\xi_2}^3$ are interchanged under the action of the group $\mathbb{Z}_2 = S(3)/A(3)$ on the ring V_1. Applying the construction used for solving quadratic equations to the polynomials $R_{\xi_1}^3$ and $R_{\xi_2}^3$, we obtain that these polynomials can be expressed through the symmetric polynomials $R_{\xi_1}^3 + R_{\xi_2}^3$ and $(R_{\xi_1}^3 - R_{\xi_2}^3)^2$. We finally obtain that the polynomial x_1 can be expressed through the symmetric polynomials $R_1 \in V_2$, $R_{\xi_1}^3 + R_{\xi_2}^3 \in V_2$, and $(R_{\xi_1}^3 - R_{\xi_2}^3)^2 \in V_2$ with the help of root extractions of the second and third degrees and the arithmetic operations. To write down an explicit formula for the solution, it remains only to express these symmetric polynomials through the elementary symmetric polynomials.

Equations of the Fourth Degree   The reason why equations of the fourth degree are solvable is that the group $S(4)$ is solvable. The group $S(4)$ is solvable because there exists a homomorphism $\pi : S(4) \to S(3)$ whose kernel is the Abelian group $Kl = \mathbb{Z}_2 \oplus \mathbb{Z}_2$. The homomorphism π can be described in the following way. There exist exactly three ways to split a four-element set into pairs of elements. Every permutation of the four elements gives rise to a permutation of these splittings. This correspondence defines the homomorphism π. The kernel Kl of this homomorphism is a normal subgroup of the group $S(4)$ consisting of four permutations: the identity permutation and the three permutations, each of which is a product of two disjoint transpositions.

Assume that the characteristic of a field K is not equal to 2 and that K contains the three cube roots of unity. The group $S(4)$ acts on the polynomial ring

$K[x_1, x_2, x_3, x_4] = V$. Denote by V_1 the invariant subalgebra of the normal subgroup Kl of the group $S(4)$. Thus the polynomial ring $V = K[x_1, x_2, x_3, x_4]$ carries an action of the Abelian group Kl with the invariant subalgebra V_1. On the ring V_1, there is an action of the solvable group $S(3) = S(4)/\mathrm{Kl}$, and the invariant subalgebra with respect to this action is the ring V_2 of symmetric polynomials.

Let A and B be operators corresponding to the permutations x_2, x_1, x_4, x_3 and x_3, x_4, x_1, x_2 of the variables x_1, x_2, x_3, x_4. The operators A and B generate the group Kl. The following identities hold: $A^2 = B^2 = E$. The roots of the polynomial $T(t) = t^2 - 1$ annihilating the operators A and B are equal to $+1$, -1. The group Kl is the sum of two copies of the group with two elements, the first copy being generated by A, and the second copy by B.

The element x_1 has four Lagrange resolvents with respect to the action of commuting operators A and B generating the group Kl:

$$R_{1,1} = \frac{1}{4}(x_1 + x_2 + x_3 + x_4),$$

$$R_{-1,1} = \frac{1}{4}(x_1 - x_2 + x_3 - x_4),$$

$$R_{1,-1} = \frac{1}{4}(x_1 + x_2 - x_3 - x_4),$$

$$R_{-1,-1} = \frac{1}{4}(x_1 - x_2 - x_3 + x_4).$$

The element x is equal to the sum of these resolvents, $x_1 = R_{1,1} + R_{-1,1} + R_{1,-1} + R_{-1,1}$, and the squares $R_{1,1}^2$, $R_{-1,1}^2$, $R_{1,-1}^2$, $R_{-1,1}^2$ of the Lagrange resolvents belong to the algebra V_1. Therefore, x_1 is expressible through the elements of the algebra V_1 with the help of the arithmetic operations and extraction of square roots. In turn, the elements of the algebra V_1 can be expressed through symmetric polynomials, since this algebra carries an action of the group $S(3)$ with the invariant subalgebra V_2 (see the solution of cubic equations above).

Let us show that this argument provides an explicit reduction of a fourth-degree equation to a cubic equation. Indeed, the resolvent $R_{1,1} = \frac{1}{4}\sigma_1$ is a symmetric polynomial, and the squares of the resolvents $R_{-1,1}$, $R_{1,-1}$, and $R_{-1,1}$ are permuted under the action of the group $S(4)$ (see the description of the homomorphism $\pi : S(4) \to S(3)$ above). Since the elements $R_{-1,1}^2$, $R_{1,-1}^2$, and $R_{-1,1}^2$ are only being permuted, the elementary symmetric polynomials of them are invariant under the action of the group $S(4)$ and hence belong to the ring V_2. Thus the polynomials

$$b_1 = R_{-1,1}^2 + R_{1,-1}^2 + R_{-1,1}^2,$$

$$b_2 = R_{-1,1}^2 R_{1,-1}^2 + R_{1,-1}^2 R_{-1,-1}^2 + R_{-1,-1}^2 R_{-1,1}^2,$$

$$b_3 = R_{-1,1}^2 R_{1,-1}^2 R_{-1,-1}^2,$$

are symmetric polynomials in x_1, x_2, x_3, and x_4, therefore, b_1, b_2, and b_3 are expressible explicitly through the coefficients of the equation

$$x^4 + a_1 x^3 + a_2 x^2 + a_3 x + a_4 = 0, \tag{1.1}$$

whose roots are x_1, x_2, x_3, x_4. To solve Eq. (1.1), it suffices to solve the equation

$$r^3 - b_1 r^2 + b_2 r - b_3 = 0 \tag{1.2}$$

and set $x = \frac{1}{4}(-a_1 + \sqrt{r_1} + \sqrt{r_2} + \sqrt{r_3})$, where r_1, r_2, and r_3 are roots of Eq. (1.2).

We conclude this section by giving yet another beautiful explicit reduction of a fourth-degree equation to a third-degree equation based on consideration of a pencil of plane quadrics [Berger 87].

The coordinates of the intersection points of two plane conics $P = 0$ and $Q = 0$, where P and Q are given second-degree polynomials in x and y, can be found by solving one cubic and several quadratic equations. Indeed, every conic of the pencil $P + \lambda Q = 0$, where λ is an arbitrary parameter, passes through the points we are looking for. For some value λ_0 of the parameter λ, the conic $P + \lambda Q = 0$ splits into a pair of lines. This value satisfies the cubic equation $\det(\tilde{P} + \lambda \tilde{Q}) = 0$, where $\tilde{P}$ and $\tilde{Q}$ are 3×3 matrices of the quadratic forms corresponding to the equations of the conics in homogeneous coordinates. The equation of each line forming the degenerate conic $P + \lambda_0 Q = 0$ can be found by solving a quadratic equation: every such line passes through the center of the conic[2] whose coordinates can be expressed through the coefficients of the conic with the help of arithmetic operations and through one of the intersection points of the conic with any fixed line. To find the coordinates of this point, one needs to solve a quadratic equation. An equation of the line passing through two given points can be found with the help of arithmetic operations. If the equations of the lines into which the conic $P + \lambda_0 Q = 0$ splits are known, then to find the desired points, it remains only to solve the quadratic equations on the intersection points of the conic $P = 0$ and each of the two lines constituting the degenerate conic.

Therefore, the general equation of the fourth degree reduces to a cubic equation with the help of arithmetic operations and quadratic root extractions. Indeed, the roots of an equation $a_0 x^4 + a_1 x^3 + a_2 x^2 + a_3 x + a_4 = 0$ are projections onto the x-axis of the intersection points of the conics $y = x^2$ and $a_0 y^2 + a_1 xy + a_2 y + a_3 x + a_4 = 0$.

1.2 Fixed Points under an Action of a Finite Group and Its Subgroups

Here, we prove one of the central theorems of Galois theory, according to which distinct subgroups in a finite group of field automorphisms have distinct invariant subfields. From this point until Sect. 1.11, we shall assume that all the fields

---

[2]The center of the conic is the point p in the plane such that $\frac{\partial P}{\partial x} = \frac{\partial P}{\partial y} = 0$ (*translator's note*).

under consideration are infinite. (Galois theory holds for finite fields as well; see Sect. 1.11.) The proof is based on a simple explicit construction using the Lagrange interpolating polynomial and on a geometrically obvious statement that a vector space cannot be covered by a finite number of proper vector subspaces.

We start with the geometric statement. Let V be an affine space (possibly infinite-dimensional) over some infinite field.

Proposition 1.2.1 *The space V cannot be represented as a union of a finite number of its proper affine subspaces.*

Proof We use induction on the number of affine subspaces. Suppose that the statement has been proved for every union of fewer than n proper affine subspaces. Suppose that the space V is representable as a union of n proper affine subspaces $V_1, \ldots, V_n$. Consider any affine hyperplane $\hat{V}$ in the space V containing the first of these subspaces, V_1. The space V is a union of an infinite family of disjoint affine hyperplanes parallel to $\hat{V}$. At most n hyperplanes from this family contain one of the subspaces $V_1, \ldots, V_n$. Take any other hyperplane from the family. To this hyperplane and its intersections with the affine subspaces $V_2, \ldots, V_n$ the induction hypothesis applies, which concludes the proof. $\square$

Corollary 1.2.2 *Suppose that a finite group of linear transformations acts on a vector space V over an infinite field. Then there exists a vector a such that the restriction of the action to the orbit of a is free.*

Proof The fixed-point set of a linear transformation is a vector subspace. If the linear transformation is different from the identity, then this subspace is proper. We can choose a to be any vector not belonging to the union of the fixed-point subspaces of nontrivial transformations from the group. $\square$

The stabilizer $G_a \subset G$ of a vector $a \in V$ is defined as the subgroup consisting of all elements $g \in G$ that fix the vector a, i.e., such that $g(a) = a$.

In general, not every subgroup G_0 of a finite linear group G is the stabilizer of some vector a. As an example, consider the cyclic group of linear transformations of the complex line generated by the multiplication by a primitive nth root of unity. If the number n is not prime, then this cyclic group has a nontrivial cyclic subgroup, but the stabilizers of all vectors are trivial (the identity subgroup for every element $a \neq 0$, and the entire group for $a = 0$). Thus the existence of a vector a stable only under the action of the subgroup G_0 is not obvious, and is not true in general, i.e., for all representations of the group G.

Lemma 1.2.3 *Let G_a and G_b be stabilizers of vectors a and b in some vector space V. Then the subspace L spanned by the vectors a and b contains a vector c whose stabilizer G_c is equal to $G_a \cap G_b$.*

Proof The subgroup $G_a \cap G_b$ fixes all vectors of the space L. However, every element $g \notin G_a \cap G_b$ acts nontrivially either on the element a or on the element b.

Vectors from L stable under the action of a fixed element $g \notin G_a \cap G_b$ form a proper subspace in L. By Proposition 1.2.1, such subspaces cannot cover the entire space L. □

Let G be a group of automorphisms of a field P. Fixed elements under the action of the group G form a subfield, which we denote by K. The field P can be viewed as a vector space over the field K.

In Galois theory, the following theorem plays a major role:

Theorem 1.2.4 *Let G be a finite group of automorphisms of a field P. Then for every subgroup G_0 of the group G there exists an element $x \in P$ whose stabilizer coincides with the subgroup G_0.*

In the proof of this theorem, it will be convenient to use the space $P[t]$ of polynomials with coefficients from the field P. Every element f of the space $P[t]$ has the form $f = a_0 + \cdots + a_m t^m$, where $a_0, \ldots, a_m \in P$. A polynomial $f \in P[t]$ defines a map $f : P \to P$ taking a point $x \in P$ to the point $f(x) = a_0 + \cdots + a_m x^m$. Every automorphism σ of the field P gives rise to the induced automorphism of the ring $P[t]$ mapping a polynomial $f = a_0 + \cdots + a_m t^m$ to the polynomial $f^\sigma = \sigma(a_0) + \cdots + \sigma(a_m) t^m$. For every element $x \in P$, the following identity holds: $f^\sigma(\sigma x) = \sigma(f(x))$. Thus the automorphism group of the field P acts on the ring $P[t]$. For every $k \geq 0$, the space $P_k[t]$ of polynomials of degree $\leq k$ is invariant under this action.

Lemma 1.2.5 *Suppose that a group G of automorphisms of the field P contains m elements. Then for every subgroup G_0 of the group G, there exists a polynomial f whose degree is less than m and whose stabilizer coincides with the group G_0.*

Proof Indeed, by Corollary 1.2.2, there exists an element $a \in P$ on whose orbit O the action of the group G is free. In particular, the orbit O contains exactly m elements. Suppose that the subgroup G_0 contains k elements. Then the group G has $q = m/k$ right G_0-cosets. Under the action of the subgroup G_0, the set O splits into q orbits O_j, $j = 1, \ldots, q$. Fix q distinct elements $b_1, \ldots, b_q$ in the invariant field K, and form the Lagrange polynomial of degree less than m that takes the value b_j at every element of the subset O_j, $j = 1, \ldots, q$. The polynomial f satisfies the assumptions of the lemma.

Indeed, the polynomial f is invariant under an automorphism σ if and only if for every element x of the field P, the equality $f(\sigma(x)) = \sigma(f(x))$ holds. Since the polynomial f has degree less than m and the set O contains m elements, it suffices to verify the equality at all elements of the set O. By construction of the polynomial f, the equality $f(\sigma(x)) = \sigma(f(x))$ holds if and only if $\sigma \in G_0$. □

We now proceed with the proof of the theorem. Consider a polynomial $f(x) = a_0 + \cdots + a_{m-1} x^{m-1}$ whose stabilizer is equal to G_0. The intersection of the stabilizers of the coefficients $a_0, \ldots, a_{m-1}$ of this polynomial coincides with the subgroup G_0. Consider the vector subspace L over the invariant subfield $K \subset P$ with respect to the action of the group G spanned by the coefficients $a_0, \ldots, a_{m-1}$. By Lemma 1.2.3, there exists a vector $c \in L$ whose stabilizer is equal to G_0.

1.3 Field Automorphisms and Relations Between Elements in a Field

In this section, we consider a finite group of field automorphisms. We prove the following two theorems from Galois theory.

The first theorem (Sect. 1.3.2) states that every element of a field is algebraic over the invariant subfield of a group of field automorphisms. Moreover, every element satisfies a separable equation (see Sect. 1.3.1) over the invariant subfield.

Suppose that y and z are two elements of the field. Under what conditions does there exist a polynomial T with coefficients from the invariant subfield such that $z = T(y)$? The second theorem (Sect. 1.3.4) states that such a polynomial T exists if and only if the stabilizer of the element y lies in the stabilizer of the element z.

1.3.1 Separable Equations

Let $T(t)$ be a polynomial over a field K, $T'(t)$ its derivative, and $D(t)$ the greatest common divisor of these polynomials.

Proposition 1.3.1 *A root of the polynomial T of multiplicity $k > 1$ is also a root of the polynomial D of multiplicity $\geq (k - 1)$.*

Proof Suppose that $T(t) = (t - x)^k Q(t)$. Then

$$T'(t) = k(t - x)^{k-1} Q(t) + (t - x)^k Q'(t)$$

(if k is divisible by the characteristic of the field, then the first summand vanishes). $\square$

A polynomial T over a field K is called *separable* if every root of T (in general, the roots lie in some extension of K) is simple. If T is separable, we also call the equation $T = 0$ separable, and if T is irreducible, we call the equation $T = 0$ irreducible as well. Proposition 1.3.1 implies that every irreducible polynomial over a field of zero characteristic is separable: indeed, an irreducible polynomial of degree n cannot have a common factor with its derivative, which has degree $n - 1$ (in a field of positive characteristic, the derivative of a nonconstant polynomial can be identically equal to zero). In the next subsection, we will see that only separable equations are important for Galois theory. For inseparable equations, Galois theory does not exist in full generality. It is not hard to prove that every irreducible equation over a finite field is separable. Let us give an example of an inseparable equation over an infinite field. If the equation $x^p = a$ has a root in a field K of characteristic p, then the multiplicity of this root is equal to p: indeed, if $x_0^p = a$, then $x^p - a = (x - x_0)^p$. Hence, if the equation $x^p = a$ does not have a root in the field K, then this equation is irreducible and inseparable.

Example 1.3.2   Let $K = \mathbb{Z}/p\mathbb{Z}(t)$ be the field of rational functions over the finite field of p elements. The equation $x^p = t$ is irreducible and inseparable.

1.3.2 Algebraicity over the Invariant Subfield

Let P be a commutative algebra with no zero divisors on which a group π acts by automorphisms, and let K be the invariant subalgebra. We do not assume that the group π is finite (although for Galois theory, it suffices to consider the actions of finite groups).

Theorem 1.3.3

1. *The stabilizer of an element* $y \in P$ *algebraic over* K *has finite index in the group* π.
2. *If the stabilizer of an element* $y \in P$ *has finite index n in the group* π, *then y is a root of an irreducible separable polynomial over K of degree n whose leading coefficient is equal to one.*

Proof

1. Suppose that an element y satisfies an algebraic equation

$$p_0 y^n + \cdots + p_n = 0 \tag{1.3}$$

   with coefficients p_i from the invariant subalgebra K. Then every automorphism of the algebra P that fixes all elements of K maps the element y to one of the roots of Eq. (1.3). There are no more than n of these roots, and hence the index of the stabilizer of y in the group π does not exceed n.
2. Suppose that the stabilizer G of an element y has index n in the group π. Then the orbit of the element y under the action of the group π contains precisely n distinct elements. Let $y_1, \ldots, y_n$ denote the elements of this orbit. Consider the polynomial $Q(y) = (y - y_1) \cdots (y - y_n)$. Under the permutation of the points $y_1, \ldots, y_n$, the factors of the polynomial Q get permuted, but the polynomial itself does not change. Hence, the coefficients of Q belong to the algebra of invariants K. The element y satisfies the algebraic equation $Q(y) = 0$ over the algebra K. This equation is irreducible (i.e., the polynomial Q does not admit a factorization into two polynomials of positive degree whose coefficients lie in the algebra K). Indeed, if it is reducible, then y satisfies an algebraic equation of smaller degree over the algebra K, and the orbit of y contains fewer than n elements. All roots of the polynomial Q are distinct; hence the polynomial is separable. $\square$

Remark 1.3.4 Viète's formulas allow one to find the coefficients of the polynomial Q. The elementary symmetric functions $\sigma_1 = y_1 + \cdots + y_n$, $\sigma_2 = \sum_{i<j} y_i y_j$, $\ldots, \sigma_n = y_1 \cdots y_n$ of the orbit points $y_1, \ldots, y_n$ belong to the invariant subalgebra K, and

$$Q(y) = y^n - \sigma_1 y^{n-1} + \sigma_2 y^{n-2} + \cdots + (-1)^n \sigma_n = 0.$$

1.3.3 Subalgebra Containing the Coefficients of the Lagrange Polynomial

In this subsection, we consider the Lagrange polynomial constructed by a special data set, and an estimate is made on the subalgebra containing its coefficients. These results will be used in Sect. 1.3.4.

Let P be a commutative algebra without zero divisors, suppose $y_1, \ldots, y_n$ are distinct elements of the algebra P, and $Q \in P[y]$ is a polynomial of degree n with leading coefficient 1 that vanishes at the points $y_1, \ldots, y_n$, i.e.,

$$Q(y) = (y - y_1) \cdots (y - y_n).$$

Consider the following problem.

Problem 1.3.5 For given elements $z_1, \ldots, z_n$ of the algebra P, find the Lagrange polynomial T taking the values $z_i Q'(y_i)$ at points y_i.

Let Q_i denote the polynomial $Q_i(y) = \prod_{j \neq i}(y - y_i)$. The following statement is obvious.

Proposition 1.3.6 *The desired Lagrange polynomial T is equal to $\sum_{i=1}^{n} z_i Q_i(y)$.*

Let us formulate a more general statement (related to this problem), which we will not use and will not prove.

Proposition 1.3.7 *Suppose that a subalgebra K of the algebra P contains the coefficients of the polynomial Q and the elements $m_0, \ldots, m_{n-1}$, where $m_k = \sum z_i y_i^k$. Then the coefficients of the polynomial T belong to the subalgebra K, i.e., $T \in K[y]$.*

We will need the following special case of this statement. Let π be a group of automorphisms of the algebra P, $K \subset P$ the algebra of invariants under the action of π, Y the orbit of an element $y_1 \in P$ under the group action. Let $f : Y \to P$ be a map commuting with the action of the group π, that is, $f \circ g = g \circ f$ if $g \in \pi$. Put $z_1 = f(y_1), \ldots, z_n = f(y_n)$.

Proposition 1.3.8   *The coefficients of the polynomial T defined in Proposition* 1.3.6 *belong to the algebra of invariants K.*

Proof The action of elements g of the group π permutes the summands of the polynomial T: if $g(y_i) = y_j$, then g maps the polynomial $z_i Q_i(y)$ to the polynomial $z_j Q_j(y)$. Hence, the polynomial T does not change under the action of the group π, that is, $T \in K[y]$. □

1.3.4   Representability of One Element Through Another Element over the Invariant Field

Let P be a field on which a group π of automorphisms acts, and K the corresponding invariant field. Suppose that x and y are elements of the field P algebraic over the field K, and G_y, G_z are their stabilizers. By Theorem 1.3.3, the element y (the element z) is algebraic over the field K if and only if the group G_y (respectively the group G_z) has finite index in the group π. Under what conditions does z belong to the extension $K(y)$ of the field K obtained by adjoining the element y? The answer to this question is provided by the following theorem:

Theorem 1.3.9   *The element z belongs to the field $K(y)$ if and only if the stabilizer G_z of the element z includes the stabilizer G_y of the element y.*

Proof In one direction, the theorem is obvious: every element of the field $K(y)$ is fixed under the action of the group G_y. In other words, the stabilizer of every element of $K(y)$ contains the group G_y.

We will prove the converse statement in a stronger form. We will now assume that P is a commutative algebra with no zero divisors (which is not necessarily a field), π is a group of automorphisms of the algebra P, K is the invariant subalgebra, y and z are elements of the algebra P whose stabilizers G_y and G_z have finite index in the group π. Let Q denote an irreducible monic polynomial over the algebra K such that $Q(y) = 0$ (see part 2 of Theorem 1.3.3).

Proposition 1.3.10   *If $G_z \supseteq G_y$, then there exists a polynomial T with coefficients in the algebra K for which we have $zQ'(y) = T(y)$.*

Proof Let S denote the set of right G_y-cosets in the group π. Suppose that the set S consists of n elements. Number the elements $s_1, \ldots, s_n$ of this set in such a way that the coset of π containing the identity element has index 1. Let g_i be any representative of the coset s_i in the group π. The images $g_i(y)$, $g_i(z)$ of the elements y, z under the action of an automorphism g_i do not depend on the choice of a representative g_i in the class s_i. We denote these images by y_i and z_i, respectively. All elements $y_1, \ldots, y_n$ are distinct by construction, whereas some of the elements $z_1, \ldots, z_n$ may coincide. For every nonnegative integer k, the element $m_k = z_1 y_1^k +$

$\cdots + z_n y_1^k$ is invariant under the action of the group π, and hence belongs to the invariant subalgebra K. To conclude the proof, it remains to use Proposition 1.3.7. Proposition 1.3.10 and Theorem 1.3.9 are thus proved.   □

1.4   Action of a k-Solvable Group and Representability by k-Radicals

In this section, we consider a field P with an action of a finite group G of automorphisms and the invariant subfield K. As before, we assume that the order n of the group G is not divisible by the characteristic of the field P and that the field P contains all nth roots of unity. Note that now n is the order of G and not the number of links in a chain of subgroups (see the definition below). Let us now denote the number of links in such a chain by m. A definition of a k-solvable group will be given. We will prove that if the group G is k-solvable, then every element of the field P can be expressed through the elements of the field K by radicals and solutions of auxiliary algebraic equations of degree at most k. The proof is based on the theorems given in preceding sections.

Definition 1.4.1 A group G is called k-*solvable* if it has a chain of nested subgroups

$$G = G_0 \supset G_1 \supset \cdots \supset G_m = e$$

such that for each i, $0 < i \leq m$, either the index of the subgroup G_i in the group G_{i-1} does not exceed k, or G_i is a normal subgroup of the group G_i and the quotient group G_{i-1}/G_i is Abelian.

Theorem 1.4.2 *Let G be a finite k-solvable group of order n acting by automorphisms of a field P containing all nth roots of unity. Then every element x of the field P can be expressed through the elements of the invariant subfield K with the help of arithmetic operations, root extractions, and solving auxiliary algebraic equations of degree k or less.*

Proof Let $G = G_0 \supset G_1 \supset \cdots \supset G_m = e$ be a chain of nested subgroups satisfying the assumptions in the definition of a k-solvable group. Denote by $K = K_0 \subset \cdots \subset K_m = P$ the chain of invariant subfields corresponding to the actions of the groups $G_0, \ldots, G_m$.

Suppose that the group G_i is a normal subgroup of the group G_{i-1} and that the quotient G_{i-1}/G_i is Abelian. The Abelian quotient group G_{i-1}/G_i acts on the invariant subfield K_i leaving the invariant subfield K_{i-1} pointwise fixed. Therefore, every element of the field K_i is expressible through the elements of the subfield K_{i-1} by means of summation and root extraction (see Theorem 1.1.3 from Sect. 1.1.1).

Suppose that the group G_i is a subgroup of index $m \leq k$ in the group G_{i-1}. There exists an element $a \in P$ whose stabilizer equals G_i (Theorem 1.2.4). The field K_i

carries an action of the group G_{i-1} of automorphisms with the invariant sub-field K_{i-1}. Since the index of the stabilizer G_i of the element a in the group G is equal to m, the element a satisfies an algebraic equation of degree $m \leq k$ over the field K_{i-1}. By Theorem 1.3.9, every element of the field K_i is a polynomial in a with coefficients from the field K_{i-1}.

Repeating the same argument, we will be able to express every element of the field P through the elements of the field K with the help of the arithmetic operations, root extractions, and solutions of auxiliary algebraic equations of degree k or less. $\square$

1.5 Galois Equations

A separable algebraic equation over a field K is called a *Galois equation* if the extension of the field K obtained by adjunction of any single root of this equation to K contains all other roots. In this section, we prove that for every separable algebraic equation over the field K there exists a Galois equation for which the extension of the field K obtained by adjunction of all roots of the initial equation coincides with the extension obtained by adjunction of a single root of the Galois equation. The proof is based on Theorem 1.3.9 from Sect. 1.3.4. Galois equations are convenient tools for constructing Galois groups (see Sects. 1.6 and 1.7).

Let K be any field. Denote by $\tilde{P}$ the algebra $K[x_1, \ldots, x_m]$ of polynomials over the field K in the variables $x_1, \ldots, x_m$. The algebra $\tilde{P}$ carries an action of a group π of automorphisms isomorphic to the permutation group $S(m)$ on m elements: the action of the group consists in the simultaneous permutation of the variables in all polynomials from the ring $K[x_1, \ldots, x_m]$. The invariant subalgebra $\tilde{K}$ with respect to this action consists of all symmetric polynomials in the variables $x_1, \ldots, x_m$.

Let $y \in \tilde{P}$ be some polynomial in m variables whose orbit under the action of the group $S(m)$ contains exactly $n = m!$ distinct elements $y = y_1, \ldots, y_n$. Let Q denote a polynomial over the algebra $\tilde{K}$ whose roots are the elements $y_1, \ldots, y_n \in \tilde{P}$ (see Theorem 1.3.3). The derivative of the polynomial Q does not vanish at its roots $y_1, \ldots, y_n$. Applying Proposition 1.3.10 to the action of the group $S(m)$ on the algebra $\tilde{P}$ with the invariant subalgebra $\tilde{K}$, we obtain the following corollary:

Corollary 1.5.1 *For every element $F \in \tilde{P} = K[x_1, \ldots, x_m]$, there exists a polynomial T whose coefficients are symmetric polynomials in the variables $x_1, \ldots, x_m$ such that the following identity holds:*

$$F Q'(y) = T(y).$$

Let $b_0 + b_1 x + \cdots + b_m x^m = 0$ be an algebraic equation over the field K, $b_i \in K$, whose roots $x_1^0, \ldots, x_m^0$ are distinct. Let P be the field obtained from K by adjoining all these roots. Consider the map $\pi : K[x_1, \ldots, x_m] \to P$, assigning to each polynomial its value at the point $(x_1^0, \ldots, x_m^0) \in P^m$.

Corollary 1.5.2 *Let* $y \in K[x_1, \ldots, x_m]$ *be a polynomial such that all* $n = m!$ *polynomials obtained from* y *by all possible permutations of the variables assume distinct values at the point* $(x_1^0, \ldots, x_m^0) \in P^m$. *Then the value of the polynomial* y *at this point generates the field* P *over the field* K.

Proof Indeed, the algebraic elements $x_1^0, \ldots, x_m^0$ generate the field P over the field K. Therefore, every element of the field P is the value of some polynomial from the ring $K[x_1, \ldots, x_m]$ at the point $(x_1^0, \ldots, x_m^0)$. However, by Corollary 1.5.1, every polynomial F multiplied by $Q'(y)$ is representable as a polynomial T of y with coefficients from the algebra $\tilde{K}$. We plug in the point $(x_1^0, \ldots, x_m^0)$ into the corresponding identity $F(x_1, \ldots, x_m) Q'(y) = T(y)$. By our assumption, all $n = m!$ roots of the polynomial Q assume distinct values at the point $(x_1^0, \ldots, x_m^0)$. Therefore, the function $Q'(y)$ is different from 0 at this point, and the values of all symmetric polynomials at the point $(x_1^0, \ldots, x_m^0)$ belong to the field K (since symmetric polynomials of the roots of an equation can be expressed through the coefficients of this equation). □

Lemma 1.5.3 *For any distinct elements* $x_1^0, \ldots, x_m^0$ *of the field* $P \supseteq K$ *there exists a linear polynomial* $y = \lambda_1 x_1 + \cdots + \lambda_m x_m$ *with coefficients* $\lambda_1, \ldots, \lambda_m$ *from the field* K *such that all* $n = m!$ *polynomials obtained from* y *by permutations of the variables assume distinct values at the point* $(x_1^0, \ldots, x_m^0) \in P^m$.

Proof Consider the $n = m!$ points obtained from the point $(x_1^0, \ldots, x_m^0)$ by all possible permutations of the coordinates. For every pair of points, the linear polynomials assuming the same values at these points form a proper vector subspace in the vector space of all linear polynomials with coefficients from the field K. The proper subspaces corresponding to pairs of points cannot cover the entire space (Proposition 1.2.1). Every linear polynomial y not lying in the union of the subspaces described above has the desired property. □

Definition 1.5.4 An equation $a_0 + a_1 x + \cdots + a_m x^m = 0$ over the field K is called a *Galois equation* if its roots $x_1^0, \ldots, x_m^0$ have the following property: for every pair of roots x_i^0, x_j^0, there exists a polynomial $P_{i,j}(t)$ over the field K such that $P_{i,j}(x_i^0) = x_j^0$.

Theorem 1.5.5 *Suppose that a field* P *is obtained from the field* K *by adjoining all roots of an algebraic equation over the field* K *with no multiple roots. Then the same field* P *can be obtained from the field* K *by adjoining a single root of some (in general, distinct) irreducible Galois equation over the field* K.

Proof By the assumption of the theorem, all roots $x_1^0, \ldots, x_m^0$ of the equation are distinct. Consider a linear homogeneous polynomial y with coefficients in the field K such that all $n = m!$ linear polynomials obtained from y by permutations of the variables assume distinct values at the point $(x_1^0, \ldots, x_m^0)$. Consider an equation of degree n over the field K whose roots are these values. By the corollary proved above, the equation thus obtained is a Galois equation, and its roots generate the field P. The Galois equation we have obtained may turn out to be reducible. Equating any irreducible component of it to zero, we obtain a desired Galois equation. □

1.6 Automorphisms Connected with a Galois Equation

In this section, we construct a group of automorphisms of an extension obtained from the base field by adjoining all roots of some Galois equation. We will show (Theorem 1.6.2) that the invariant subfield of this group coincides with the field of coefficients.

Let $Q = b_0 + b_1 x + \cdots + b_n x^n$ be an irreducible polynomial over the field K. Then all fields generated over the field K by a single root of the polynomial Q are isomorphic to each other and admit the following abstract description: every such field is isomorphic to the quotient of the ring $K[x]$ by the ideal I_Q generated by the irreducible polynomial Q. We denote this field by $K[x]/I_Q$.

Let M be an extension of the field K containing all n roots $x_1^0, \ldots, x_n^0$ of the equation $Q(x) = 0$. With every root x_i^0, we associate the field K_i obtained by adjoining the root x_i^0 to the field K. All the fields K_i, $i = 1, \ldots, n$, are isomorphic to each other and are isomorphic to the field $K[x]/I_Q$. Denote by σ_i the isomorphic map of the field $K[x]/I_Q$ to the field K_i that fixes all elements of the coefficient field K and takes the polynomial x to the element x_i^0.

Lemma 1.6.1 *Suppose that the equation $Q = b_0 + b_1 x + \cdots + b_n x^n = 0$ is irreducible over the field K. Then the images $\sigma_i(a)$ of an element a of the field $K[x]/I_Q$ in the field M under all isomorphisms σ_i, $i = 1, \ldots, n$, coincide if and only if the element a lies in the coefficient field K.*

Proof If $b = \sigma_1(a) = \cdots = \sigma_n(a)$, then the element b is equal to $b = (\sigma_1(a) + \cdots + \sigma_n(a))n^{-1}$ (in the case under consideration, $n \neq 0$ in the field K). Therefore, the element b is the value of a symmetric polynomial in the roots $x_1^0, \ldots, x_m^0$ of the equation $Q(x) = 0$; hence it belongs to the field K. □

We are now ready for the main theorem of this section:

Theorem 1.6.2 *Suppose that a field P is obtained from the field K by adjoining all roots of an irreducible algebraic equation over the field K. Then an element $b \in P$ is fixed by all automorphisms of P fixing all elements of K if and only if $b \in K$.*

Proof By Theorem 1.5.5 we can assume that the field P is obtained from the field K by adjoining all roots (or, which is the same, a single root) of some irreducible Galois equation. By definition of a Galois equation, all the fields K_i mentioned in Lemma 1.6.1 coincide with the field P. The isomorphism $\sigma_j \sigma_i^{-1}$ between the field K_i and the field K_j is an automorphism of the field P fixing all elements of the field K. By the lemma, an element b is fixed under all such automorphisms if and only if $b \in K$. □

1.7 The Fundamental Theorem of Galois Theory

In Sects. 1.2, 1.3, 1.5, and 1.6 we in fact proved the central theorems of Galois theory. In this section, we give a summary. We define Galois extensions (Sect. 1.7.1) and Galois groups (Sect. 1.7.2), we prove the fundamental theorem of Galois theory (Sect. 1.7.3), and we discuss the properties of the Galois correspondence (Sect. 1.7.4) and the behavior of the Galois group under extensions of the coefficient field.

1.7.1 Galois Extensions

We give two equivalent definitions:

Definition 1.7.1 A field P obtained from a field K by adjunction of all roots of a separable algebraic equation over the field K is called a *Galois extension* of the field K.

Definition 1.7.2 A field P is a *Galois extension* of its subfield K if there exists a finite group G of automorphisms of the field P whose invariant subfield is the field K.

Proposition 1.7.3 *Definitions* 1.7.1 *and* 1.7.2 *are equivalent. The group G from Definition* 1.7.2 *coincides with the group of all automorphisms of the field P over the field K. It follows that the group G is uniquely defined.*

Proof If the field P is a Galois extension of the field K in the sense of Definition 1.7.1, then by Theorem 1.6.2, the field P is also a Galois extension of the field K in the sense of Definition 1.7.2. Suppose now that the field P is a Galois extension of the field K in the sense of Definition 2. By Corollary 1.2.2, there exists an element $a \in P$ that moves (does not stay fixed) under the action of every element of the group G. Consider the orbit O of the element a under the action of G. By Theorem 1.3.3, there exists an algebraic equation over the field K whose set of roots coincides with O. By Theorem 1.3.9, every element of the orbit, i.e., every root of this algebraic equation, generates the field P over the field K. Therefore, the field P is a Galois extension of the field K in the sense of Definition 1.7.1.

Every automorphism σ of the field P over the field K takes the element a to some element of the set O, since the set O is the set of all solutions of an algebraic equation with coefficients in the field K. Hence σ defines an element g of the group G such that $\sigma(a) = g(a)$. The automorphism σ must coincide with g, since a generates the field P over the field K. Therefore, the group G coincides with the group of all automorphisms of the field P over the field K. $\square$

1.7.2 Galois Groups

We now proceed with Galois groups, which are central objects in Galois theory. The *Galois group of a Galois extension* P of the field K (or just the Galois group of P over K) is defined as the group of all automorphisms of the field P over the field K. The *Galois group of a separable algebraic equation* over the field K is defined as the Galois group of the Galois extension P of K obtained by adjoining all roots of this algebraic equation to the field K.

Suppose that the field P is obtained by adjoining to K all roots of the equation

$$a_0 + a_1 x + \cdots + a_n x^n = 0 \tag{1.4}$$

over the field K. Every element σ from the Galois group of P over K permutes the roots of Eq. (1.4). Indeed, acting by σ on both parts of Eq. (1.4) yields

$$\sigma\left(a_0 + a_1 x + \cdots + a_n x^n\right) = a_0 + a_1 \sigma(x) + \cdots + a_n \left(\sigma(x)\right)^n = 0.$$

Thus, the Galois group of the field P over the field K admits a representation in the permutation group of the roots of Eq. (1.4). This representation is faithful: if an automorphism fixes all the roots of Eq. (1.4), then it fixes all elements of the field P and hence is trivial.

Definition 1.7.4 A *relation* between the roots of Eq. (1.4) over the field K is defined as any polynomial Q belonging to the ring $K[x_1, \ldots, x_n]$ that vanishes at the point $(x_1^0, \ldots, x_n^0)$, where $x_1^0, \ldots, x_n^0$ is the collection of all roots of Eq. (1.4).

Proposition 1.7.5 *Every automorphism of the Galois group preserves all relations over the field K between the roots of Eq. (1.4). Conversely, every permutation of the roots preserving all relations between the roots over the field K extends to an automorphism of the Galois group.*

Thus the Galois group of the field P over the field K can be identified with the group of all permutations of the roots of Eq. (1.4) that preserve all relations between the roots defined over the field K.

Proof If a permutation $\sigma \in S(n)$ corresponds to an element of the Galois group, then the polynomial σQ obtained from a relation Q by permuting the variables $x_1, \ldots, x_n$ according to σ also vanishes at the point $(x_1^0, \ldots, x_n^0)$. Conversely, suppose that a permutation σ preserves all relations between the roots over the field K. Extend the permutation σ to an automorphism of the field P over the field K. Every element of the field P is the value of some polynomial Q_1 belonging to the ring $K[x_1, \ldots, x_n]$ at the point $(x_1^0, \ldots, x_n^0)$. It is natural to define the value of the automorphism σ at this element as the value of the polynomial σQ_1 obtained from Q_1 by permuting the variables according to σ at the point $(x_1^0, \ldots, x_n^0)$. We need to verify that the automorphism σ is well defined. Let Q_2 be a different polynomial from the ring $K[x_1, \ldots, x_n]$ whose value at the point $(x_1^0, \ldots, x_n^0)$ coincides with the value of Q_1 at this point. But then the polynomial $Q_1 - Q_2$ is a relation between the roots over the field K. Therefore, the polynomial $\sigma Q_1 - \sigma Q_2$ must also vanish at the point $(x_1^0, \ldots, x_n^0)$, but this means exactly that the automorphism σ is well defined. $\qquad\square$

1.7.3 The Fundamental Theorem

Suppose that the field P is a Galois extension of a field K. Galois theory describes all intermediate fields, i.e., all fields lying in the field P and containing the field K. To every subgroup J of the Galois group of the field P over the field K we assign the subfield P_J consisting of all elements of P that are fixed under the action of J. This correspondence is called the *Galois correspondence*.

Theorem 1.7.6 (The fundamental theorem of Galois theory) *The Galois correspondence of a Galois extension is a one-to-one correspondence between all subgroups in the Galois group and all intermediate fields.*

Proof Firstly, by Theorem 1.2.4, distinct subgroups in the Galois group have distinct invariant subfields. Secondly, if a field P is a Galois extension of the field K, then it is also a Galois extension of every intermediate field. This is obvious if we use Definition 1.7.1 of a Galois extension. From Definition 1.7.2 of a Galois extension, it can be seen that every intermediate field is the invariant subfield for some group of automorphisms of the field P over the field K. The theorem is proved. $\qquad\square$

1.7.4 Properties of the Galois Correspondence

We now discuss the simplest properties of the Galois correspondence.

Proposition 1.7.7 *An intermediate field is a Galois extension of the coefficient field if and only if under the Galois correspondence, this field maps to a normal subgroup of the Galois group. The Galois group of an intermediate Galois extension over the coefficient field is isomorphic to the quotient of the Galois group of the initial extension by the normal subgroup corresponding to the intermediate Galois extension.*

Proof Let H be a normal subgroup of the Galois G, and L_H an intermediate field corresponding to the subgroup H. The field L_H gets mapped to itself under the automorphisms from the group G, since the fixed-point set of a normal subgroup is invariant under the action of the group (Proposition 1.1.4). The group of automorphisms of the field L_H induced by the action of the group G is isomorphic to the quotient group G/H. The invariant subfield of this induced group of automorphisms of L_H coincides with the field K. Thus if H is a normal subgroup of the group G, then L_H is a Galois extension of the field K with Galois group G/H.

Let K_1 be an intermediate Galois extension of the field K. The field K_1 is obtained from the field K by adjoining all roots of some algebraic equation over K. Every automorphism of the Galois group G can only permute the roots of this equation, and hence maps the field K_1 to itself. Suppose that the field K_1 corresponds to a subgroup H, i.e., $K_1 = L_H$. An element g of the group G takes the field L_H to the field $L_{gHg^{-1}}$. Thus, if an intermediate Galois extension K_1 corresponds to a subgroup H, then for every element $g \in G$ we have $H = gHg^{-1}$. In other words, the subgroup H is a normal subgroup of the Galois group G. $\qquad\square$

Proposition 1.7.8 *The smallest algebraic extension of a field K containing two given Galois extensions of the field K is a Galois extension of the field K.*

Proof The smallest field P containing both Galois extensions can be constructed in the following way. Suppose that the first field is obtained from the field K by adjoining all roots of a separable polynomial Q_1, and the second field by adjoining all roots of a separable polynomial Q_2. The polynomial $Q = Q_1 Q_2 / L$, where L is the greatest common divisor of Q_1 and Q_2, does not have multiple roots. The field P can be obtained by adjoining all roots of the separable polynomial Q to the field K and therefore is a Galois extension of the field K. $\qquad\square$

Proposition 1.7.9 *The intersection of two Galois extensions is a Galois extension. The Galois group of the intersection is isomorphic to a quotient group of the Galois group of each initial Galois extension.*

Proof Let P be the smallest field containing both Galois extensions. As we have proved, P is a Galois extension of the field K. The Galois group G of the field P over the field K preserves the first as well as the second extension of K. We conclude that the intersection of the two Galois extensions is also mapped to itself under the action of the group G. Therefore, by Proposition 1.7.7, the intersection of two Galois extensions is also a Galois extension. From the same proposition it follows that the Galois group of the intersection is a quotient group of the Galois group of each initial Galois extension. $\qquad\square$

1.7.5 Change of the Coefficient Field

Let

$$a_0 + a_1 x + \cdots + a_n x^n = 0 \qquad (1.5)$$

be a separable algebraic equation over the field K, and P a Galois extension of the field K obtained from K by adjoining all roots of Eq. (1.5). Consider a bigger field $\tilde{K} \supset K$ and its Galois extension $\tilde{P}$ obtained from the field $\tilde{K}$ by adjoining all roots of Eq. (1.5). What is the relation between the Galois group of $\tilde{P}$ and the Galois group G of P over K? In other words, what happens with the Galois group of Eq. (1.5) if we change the base field (i.e., pass from the field K to the field $\tilde{K}$)?

Generally speaking, as the coefficient field gets bigger, the Galois group of the same equation gets smaller, i.e., gets replaced with some subgroup. Indeed, there may be more relations between the roots of (1.5) over the bigger field. We now give a more precise statement.

Denote by K_1 the intersection of the fields P and $\tilde{K}$. The field K_1 includes the field K and lies in the field P, i.e., we have $K \subset K_1 \subset P$. By the fundamental theorem of Galois theory, the field K_1 corresponds to a subgroup G_1 of the Galois group G.

Theorem 1.7.10 *The Galois group $\tilde{G}$ of the field $\tilde{P}$ over the field $\tilde{K}$ is isomorphic to the subgroup G_1 in the Galois group G of the field P over the field K.*

Proof The Galois group $\tilde{G}$ fixes all elements of the field K (since $K \subset \tilde{K}$) and permutes the roots of Eq. (1.5). Hence the field P is mapped to itself under all automorphisms from the group $\tilde{G}$. The fixed-point set of the induced group of automorphisms of the field P consists precisely of all elements in the field P that lie in the field $\tilde{K}$, i.e., of all elements of the field $K_1 = P \cap \tilde{K}$. Therefore, the induced group of automorphisms of the field P coincides with the subgroup G_1 of the Galois group G. It remains to show that the homomorphism of the group $\tilde{G}$ into the group G_1 described above has trivial kernel. Indeed, the kernel of this homomorphism fixes all roots of Eq. (1.5), i.e., contains only the identity element of the group $\tilde{G}$. The theorem is proved. $\square$

Suppose now that under the assumptions of the preceding theorem, the field $\tilde{K}$ is itself a Galois extension of the field K with a Galois group Γ. By Proposition 1.7.8, the field K_1 is also a Galois extension of the field K in this case. Let Γ_1 denote the Galois group of the extension K_1 of the field K.

Theorem 1.7.11 (On how the Galois group changes under a change of the coefficient field) *As the coefficient field gets replaced with its Galois extension, the Galois group G of the initial equation gets replaced with its normal subgroup G_1. The quotient group G/G_1 of the group G by this normal subgroup is isomorphic to a quotient group of the Galois group of the new coefficient field $\tilde{K}$ over the old coefficient field K.*

Proof Indeed, the group G_1 corresponds to the field $P \cap \tilde{K}$, which is a Galois extension of the field K. Hence the group G_1 is a normal subgroup of the group G, and its quotient group G/G_1 is isomorphic to the Galois group of the field K_1 over the field K. But the Galois group of the field K_1 over the field K is isomorphic to the quotient group Γ/Γ_1. The theorem is proved. $\square$

1.8 A Criterion for Solvability of Equations by Radicals

In Sects. 1.8–1.10, we will be dealing with the question of solvability of algebraic equations over a field K. For simplicity, we shall always assume that K has characteristic zero. An algebraic equation over a field K is said to be *solvable by radicals* if there exists a chain of extensions $K = K_0 \subset K_1 \subset \cdots \subset K_n$ in which every field K_{j+1} is obtained from the field K_j, $j = 0, \ldots, n-1$, by adjoining some radical, and the field K_n contains all roots of this algebraic equation. Is a given algebraic equation solvable by radicals? Galois theory was created to answer this question.

In Sect. 1.8.1, we consider the group of all nth roots of unity that lie in a given field K. In Sect. 1.8.2, we consider the Galois group of the equation $x^n = a$. In Sect. 1.8.3, we give a criterion for solvability of an algebraic equation by radicals (in terms of the Galois group of this equation).

1.8.1 Roots of Unity

Let K be a field. Let K_E^* denote the multiplicative group of all roots of unity lying in the field (i.e., $a \in K_E^*$ if and only if $a \in K$, and for some positive integer n, we have $a^n = 1$).

Proposition 1.8.1 *If there is a subgroup of the group K_E^* consisting of exactly l elements, then the equation $x^l = 1$ has exactly l solutions in the field K, and the subgroup under consideration is formed by all these solutions.*

Proof Every element in a group of order l satisfies the equation $x^l = 1$. The field contains no more than l roots of this equation, and the subgroup has exactly l elements by our assumption. □

From Proposition 1.8.1, it follows, in particular, that the group K_E^* has at most one cyclic subgroup of any given finite order.

Proposition 1.8.2 *A finite Abelian group that has at most one cyclic subgroup of any given finite order is cyclic. In particular, every finite subgroup of the group K_E^* is cyclic.*

Proof From the classification theorem for finite Abelian groups it follows that an Abelian group satisfying the assumptions of the proposition is determined by the number m of its elements up to isomorphism: if $m = p_1^{k_1} \cdots p_n^{k_n}$ is a prime decomposition of m, then $G = (\mathbb{Z}/p_1^{k_1}\mathbb{Z}) \times \cdots \times (\mathbb{Z}/p_n^{k_n}\mathbb{Z})$. Therefore (see Proposition 1.8.1) the groups of roots of unity with the given number m of elements are isomorphic to each other. But in the field of complex numbers, any group of order m consisting of roots of unity is obviously cyclic. □

A cyclic group with m elements identifies with the group of residues modulo m.

Proposition 1.8.3 *The full automorphism group of the group $\mathbb{Z}/m\mathbb{Z}$ is isomorphic to the multiplicative group of all invertible elements in the ring of residues modulo m. In particular, this automorphism group is Abelian.*

Proof An automorphism F of the group $\mathbb{Z}/m\mathbb{Z}$ is uniquely determined by the element $F(1)$, which must obviously be invertible in the multiplicative group of the ring of residues. This automorphism coincides with the multiplication by $F(1)$. $\square$

Proposition 1.8.4 *Suppose that a Galois extension P of a field K is obtained from the field K by adjoining some roots of unity. Then the Galois group of the field P over the field K is Abelian.*

Proof All roots of unity that lie in the field P form a cyclic group with respect to multiplication. A transformation in the Galois group defines an automorphism of this group and is uniquely determined by this automorphism, i.e., the Galois group embeds into the full automorphism group of a cyclic group. Proposition 1.8.4 now follows from Proposition 1.8.3. $\square$

1.8.2 The Equation $x^n = a$

Proposition 1.8.5 *Suppose that a field K contains all nth roots of unity. Then the Galois group of the equation $x^n - a = 0$ over the field K is a subgroup of the cyclic group with n elements, provided that $0 \neq a \in K$.*

Proof The group of all nth roots of unity is cyclic (see Proposition 1.8.2). Let ξ be any generator of this group. Fix any root x_0 of the equation $x^n - a = 0$. Then we can number all roots of the equation $x^n - a = 0$ with residues i modulo n by setting x_i to be $\xi^i x_0$. Suppose that a transformation g from the Galois group takes the root x_0 to the root x_i. Then $g(x_k) = g(\xi^k x_0) = \xi^{k+i} x_0 = x_{k+i}$ (recall that by our assumption, $\xi \in K$, whence $g(\xi) = \xi$), i.e., every transformation of the Galois group defines a cyclic permutation of the roots. Therefore, the Galois group embeds into the cyclic group with n elements. $\square$

Lemma 1.8.6 *The Galois group G of the equation $x^n - a = 0$ over the field K, where $0 \neq a \in K$, has an Abelian normal subgroup G_1 such that the corresponding quotient G/G_1 is Abelian. In particular, the group G is solvable.*

Proof Let P be an extension of the field K obtained by adjoining to this field all roots of the equation $x^n = a$. The ratio of any two roots of the equation $x^n = a$ is an nth root of unity. This implies that the field P contains all nth roots of unity. Let K_1 denote the extension of the field K obtained by adjoining all nth roots of unity. We have the inclusions $K \subset K_1 \subset P$. Denote by G_1 the Galois group of the equation $x^n = a$ over the field K_1. By Proposition 1.8.5, the group K_1 is Abelian.

The group G_1 is a normal subgroup of the group G, since the field K_1 is a Galois extension of the field K. The quotient group G/G_1 is Abelian, since by Proposition 1.8.4, the Galois group of the field K_1 over the field K is Abelian. $\qquad\square$

1.8.3 Solvability by Radicals

The following theorem states a criterion for solvability of algebraic equations by radicals.

Theorem 1.8.7 (A criterion for solvability of equations by radicals) *An algebraic equation in one variable over a field K of characteristic zero is solvable by radicals if and only if its Galois group is solvable.*

Proof Suppose that an equation can be solved by radicals. Solvability of the equation by radicals over a field K means the existence of a chain of extensions $K = K_0 \subset K_1 \subset \cdots \subset K_n$ in which every field K_{j+1} is obtained from the field K_j, $j = 0, 1, \ldots, n-1$, by adjoining all roots of $x^n - a$, and the field K_n contains all the roots of the initial equation. Let G_j denote the Galois group of our equation over the field K_j. Let us see what happens with the Galois group when we pass from the field K_j to the field K_{j+1}. According to Theorem 1.7.10, the group G_{j+1} is a normal subgroup of the group G_j. Moreover, the quotient G_j/G_{j+1} is simultaneously a quotient of the Galois group of the field K_{j+1} over the field K_j. Since the field K_{j+1} is obtained from the field K_j by adjoining roots of $x^n - a$, we conclude by Lemma 1.8.6 that the Galois group of the field K_{j+1} over the field K_j is solvable. (When the field K contains all roots of unity, the Galois group of the field K_{j+1} over the field K_j is Abelian.) Since all roots of the algebraic equation lie in the field K_n by our assumption, the Galois group G_n of the algebraic equation over the field K_n is trivial.

Thus, if the equation can be solved by radicals, then its Galois group admits a chain of subgroups $G = G_0 \supset G_1 \supset \cdots \supset G_n$ in which every group G_{j+1} is a normal subgroup of the group G_j with a solvable quotient G_j/G_{j+1}, and the group G_n is trivial. (If the field K contains all roots of unity, then the quotients G_j/G_{j+1} are Abelian.) Thus, if the equation is solvable by radicals, then its Galois group is solvable.

Suppose now that the Galois group G of an algebraic equation over the field K is solvable. Denote by $\tilde{K}$ the field obtained from the field K by adjoining all roots of unity. The Galois group $\tilde{G}$ of the algebraic equation over the bigger field $\tilde{K}$ is a subgroup of the Galois group G. Hence the Galois group $\tilde{G}$ is solvable. Denote by $\tilde{P}$ the field obtained from the field $\tilde{K}$ by adjoining all roots of the algebraic equation. The solvable group $\tilde{G}$ acts by automorphisms of the field $\tilde{P}$ with the invariant subfield $\tilde{K}$. By Theorem 1.1.3, every element of the field $\tilde{P}$ is expressible by radicals through the elements of the field $\tilde{K}$. By definition of the field $\tilde{K}$, every element of this field is expressible through the roots of unity and the elements of the field K. The theorem is proved. $\qquad\square$

1.9 A Criterion for Solvability of Equations by k-Radicals

We say that an algebraic equation is *solvable by k-radicals* if there exists a chain of
extensions $K = K_0 \subset K_1 \subset \cdots \subset K_n$, in which for every j, $0 < j \leq n$, either the
field K_{j+1} is obtained from the field K_j by adjoining a radical, or the field K_{j+1}
is obtained from the field K_j by adjoining a solution of an equation of degree at
most k, and the field K_n contains all roots of the initial equation. Is a given algebraic equation solvable by k-radicals? In this section, we answer this question. In
Sect. 1.9.1, we discuss the properties of k-solvable groups. In Sect. 1.9.2, we prove
a criterion for solvability by k-radicals.

Let us start with the following simple statement.

Proposition 1.9.1 *The Galois group of an equation of degree $m \leq k$ is isomorphic
to a subgroup of the group $S(k)$.*

Proof Every element of the Galois group permutes the roots of the equation, and is
uniquely determined by the permutation of the roots thus obtained. Hence the Galois
group of a degree-m equation is isomorphic to a subgroup of the group $S(m)$. For
$m \leq k$, the group $S(m)$ is a subgroup of the group $S(k)$.   □

1.9.1 Properties of k-Solvable Groups

In this subsection, we show that k-solvable groups (see Sect. 1.4) have properties
similar to those of solvable groups. We start with Lemma 1.9.2, which characterizes
subgroups in the group $S(k)$.

Lemma 1.9.2 *A group is isomorphic to a subgroup of the group $S(k)$ if and only if
it has a collection of m subgroups, $m \leq k$, such that*

1. *the intersection of these subgroups contains no nontrivial normal subgroups of
   the entire group;*
2. *the sum of indices of these subgroups does not exceed k.*

Proof Suppose that G is a subgroup of the group $S(k)$. Consider a representation of
the group G as a subgroup of permutations of a set M with k elements. Suppose that
under the action of the group G, the set M splits into m orbits. Choose a single point
x_i in every orbit. The collection of stabilizers of points x_i satisfies the conditions of
the lemma.

Conversely, let a group G have a collection of subgroups $G_1, \ldots, G_m$ satisfying the conditions of the lemma. Denote by P the union of the sets P_i, where
$P_i = G/G_i$ consists of all right cosets with respect to the subgroup G_i, $1 \leq i \leq n$.
The group G acts naturally on the set P. The representation of the group G in the
group $S(P)$ of all permutations of P is faithful, since the kernel of this representation lies in the intersection of the groups G_i. The group $S(P)$ embeds into the

group $S(k)$, since the number of elements in the set P is the sum of the indices of the subgroups G_i. $\qquad\square$

Corollary 1.9.3 *Every quotient group of the symmetric group $S(k)$ is isomorphic to a subgroup of $S(k)$.*

Proof Suppose that a group G is isomorphic to a subgroup of the group $S(k)$, and G_i are subgroups in G satisfying the conditions of Lemma 1.9.2. Let π be an arbitrary homomorphism of the group G (onto some other group). Then the collection of the subgroups $\pi(G_i)$ in the group $\pi(G)$ also satisfies the conditions of the lemma. $\quad\square$

We say that a normal subgroup H of a group G is of *depth* at most k if the group G has a subgroup G_0 of index at most k such that H is the intersection of all subgroups conjugate to G_0. We say that a group is of depth at most k if its identity subgroup is of depth at most k.

A normal tower of a group G is a nested chain of subgroups $G = G_0 \supset \cdots \supset G_n = \{e\}$ in which each group is a normal subgroup of the preceding group.

Corollary 1.9.4 *If a group G is a normal subgroup of the group $S(k)$, then the group G has a nested chain of subgroups $G = G_0 \supset \cdots \supset G_n = \{e\}$ in which the group G_n is trivial, and for every $i = 0, 1, \ldots, n-1$, the group G_{i+1} is a normal subgroup of the group G_i of depth at most k.*

Proof Let G_i be a collection of subgroups in the group G satisfying the conditions of Lemma 1.9.2. Denote by F_i the normal subgroup of the group G obtained as the intersection of all subgroups conjugate to the subgroup G_i. The chain of subgroups $\Gamma_0 = F_0, \Gamma_1 = F_0 \cap F_1, \ldots, \Gamma_m = F_0 \cap F_1 \cap \cdots \cap F_m$ satisfies the conditions of the corollary. $\qquad\square$

Lemma 1.9.5 *A group G is k-solvable if and only if it admits a normal tower of subgroups $G = G_0 \supset \cdots \supset G_n = \{e\}$ in which for every i, $0 < i \le n$, either the normal subgroup G_i has depth at most k in the group G_{i-1} or the quotient G_{i-1}/G_i is Abelian.*

Proof

1. Suppose that the group G admits a normal tower $G = G_0 \supset \cdots \supset G_n = \{e\}$ satisfying the conditions of the lemma. If, for some i, the normal subgroup G_i has depth at most k in the group G_{i-1}, then the group G_{i-1}/G_i has a chain of subgroups $G_{i-1}/G_i = \Gamma_0 \subset \cdots \supset \Gamma_n = \{e\}$, in which the index of every next group of the preceding group does not exceed k. For every such number i, we can insert the chain of subgroups $G_{i-1} = \Gamma_{0,i} \supset \cdots \supset \Gamma_{0,i}$ between G_{i-1} and G_i, where π is the canonical projection to the quotient group. We thus obtain a chain of subgroups satisfying the definition of a k-solvable group.

2. Suppose that a group G is k-solvable, and $G = G_0 \supset G_1 \supset \cdots \supset G_n = \{e\}$ is
   a chain of subgroups satisfying the assumptions listed in the definition of a k-
   solvable group. We will successively replace subgroups in the chain with smaller
   subgroups. Let i be the first number for which the group G_i is not a normal
   subgroup of the group G_{i-1} but rather a subgroup of index $\leq k$. In this case,
   the group G_{i-1} has a normal subgroup H lying in the group G_i and such that
   the group G_{i-1}/H is isomorphic to a subgroup of $S(k)$. Indeed, for H, we can
   take the intersection of all subgroups in G_{i-1} conjugate to the group G_i. We can
   now modify the chain $G = G_0 \supset G_1 \supset \cdots \supset G_n = \{e\}$ in the following way: all
   subgroups labeled by numbers less than i remain the same. Every group G_j with
   $i \leq j$ gets replaced with the group $G_j \cap H$. Apply the same procedure to the
   chain of subgroups thus obtained, and so on. Finally, we obtain a normal tower
   of subgroups satisfying the conditions of the lemma. $\Box$

Theorem 1.9.6

1. *Every subgroup and every quotient group of a k-solvable group are k-solvable.*
2. *If a group has a k-solvable normal subgroup such that the corresponding quo-*
   tient group is k-solvable, then the group is also k-solvable.

Proof The only nonobvious statement of this theorem is that about a quotient group.
It follows easily from Lemma 1.9.5. $\Box$

1.9.2 Solvability by k-Radicals

The following theorem gives a criterion for solvability by k-radicals:

Theorem 1.9.7 (A criterion for solvability of equations by k-radicals) *An algebraic*
equation over a field K of characteristic zero is solvable by k-radicals if and only if
its Galois group is k-solvable.

Proof

1. Suppose that the equation can be solved by k-radicals. We need to prove that the
   Galois group of the equation is k-solvable. This is proved in exactly the same
   way as the solvability of the Galois group of an equation solvable by radicals.
       Let $K = K_0 \subset K_1 \subset \cdots \subset K_n$ be a chain of fields that arises in the solution
   of the equation by k-radicals, and $G_0 \supset \cdots \supset G_n$ the chain of Galois groups
   of the equation over these fields. By the assumption, the field K_n contains all
   roots of the equation, and therefore, the group G_n is trivial and, in particular, is
   k-solvable. Suppose that the group G_{i+1} is k-solvable. We need to prove that the
   group G_i is also k-solvable.
       If the field K_{i+1} is obtained from the field K_i by adjoining a radical, then the
   Galois group of the field K_{i+1} over the field K_i is solvable, hence k-solvable. If

the field K_{i+1} is obtained from the field K_i by adjoining all roots of an algebraic equation of degree at most k, then the Galois group of the field K_{i+1} over the field K_i is a subgroup of the group $S(k)$ (see Proposition 1.9.1), hence is k-solvable.

By Theorem 1.1.3, the group G_{i+1} is a normal subgroup of the group G_i; moreover, the quotient group G_i/G_{i+1} is simultaneously a quotient group of the Galois group of the field K_{i+1} over the field K_i. The group G_{i+1} is solvable by the induction hypothesis. The Galois group of the field K_{i+1} over the field K_i is k-solvable, as we have just proved. Using Theorem 1.9.6, we conclude that the group G_i is k-solvable.

2. Suppose that the Galois group G of an algebraic equation over the field K is k-solvable. Let $\tilde{K}$ denote the field obtained from the field K by adjoining all roots of unity. The Galois group $\tilde{G}$ of the same equation over the bigger field $\tilde{K}$ is a subgroup of the group G. Therefore, the Galois group $\tilde{G}$ is k-solvable. Let $\tilde{P}$ denote the field obtained from the field $\tilde{K}$ by adjoining all roots of the given algebraic equation. The group $\tilde{G}$ acts by automorphisms on $\tilde{P}$ with the invariant subfield $\tilde{K}$. By Theorem 1.1.3, every element of the field $\tilde{P}$ can be expressed through the elements of the field $\tilde{K}$ by taking radicals, performing arithmetic operations, and solving algebraic equations of degree at most k. By definition of the field $\tilde{K}$, every element of this field is expressible through the elements of the field K and roots of unity. The theorem is proved. $\square$

1.9.3   Unsolvability of a Generic Degree-$(k+1 > 4)$ Equation in k-Radicals

Let K be a field. A generic algebraic equation of degree k with coefficients in the field K is an equation

$$x^k + a_1 x^{k-1} + \cdots + a_0 = 0 \tag{1.6}$$

whose coefficients are sufficiently general elements of the field K. Do there exist formulas containing radicals (k-radicals) and variables $a_1, \ldots, a_k$ that give solutions of an equation $x^k + a_1^0 x^{k-1} + \cdots + a_0^0 = 0$ as one substitutes the particular elements $a_1^0, \ldots, a_k^0$ of the field K for the variables?

This question can be formalized in the following way. A generic algebraic equation can be viewed as an equation over the field $K\{a_1, \ldots, a_k\}$ of rational functions in k independent variables $a_1, \ldots, a_k$ with coefficients in the field K (in this interpretation, the coefficients of Eq. (1.6) are the elements $a_1, \ldots, a_k$ of the field $K\{a_1, \ldots, a_k\}$). We can now ask the question on solvability of Eq. (1.6) over the field $K\{a_1, \ldots, a_k\}$ by radicals (or by k-radicals).

Let us compute the Galois group of Eq. (1.6) over the field $K\{a_1, \ldots, a_k\}$. Consider yet another copy of the field $K\{a_1, \ldots, a_k\}$ of rational functions in k variables equipped with the group $S(k)$ of automorphisms acting by permutations of the variables $x_1, \ldots, x_k$. The invariant subfield $K_S\{a_1, \ldots, a_k\}$ consists of symmetric rational functions. By the fundamental theorem of symmetric functions, this field is isomorphic to the field of rational functions of the variables $\sigma_1 = x_1 + \cdots + x_k, \ldots, \sigma_n = x_1 \cdots x_k$. Therefore the map $F(a_1) = -\sigma_1, \ldots, F(a_n) = (-1)^n \sigma_n$ extends to an isomorphism $F : K\{a_1, \ldots, a_k\} \to K_S\{x_1, \ldots, x_k\}$. Let us identify the fields $K\{a_1, \ldots, a_k\}$ and $K_S\{x_1, \ldots, x_k\}$ by the isomorphism F. From the comparison of Viète's formulas with the formulas defining the map F, it becomes clear that under this identification, the variables become the roots of Eq. (1.6), the field $K\{x_1, \ldots, x_k\}$ becomes the extension of the field $K\{a_1, \ldots, a_k\}$ by adjoining all roots of Eq. (1.6), and the automorphism group $S(k)$ becomes the Galois group of Eq. (1.6). Thus we have proved the following statement:

Proposition 1.9.8 *The Galois group of Eq. (1.6) over the field $K\{a_1, \ldots, a_k\}$ is isomorphic to the permutation group $S(k)$.*

Theorem 1.9.9 *A generic algebraic equation of degree $k + 1 > 4$ is not solvable by radicals and by solving auxiliary algebraic equations of degree k or less.*

Proof The group $S(k + 1)$ has the following normal tower of subgroups: $\{e\} \subset A(k + 1) \subset S(k + 1)$, where $A(k + 1)$ is the alternating group. For $k + 1 > 4$, the group $A(k + 1)$ is simple. The group $A(k + 1)$ is not a subgroup of the group $S(k)$, since the group $A(k + 1)$ has more elements than the group $S(k)$. Thus for $k + 1 > 4$, the group $S(k + 1)$ is not k-solvable. To conclude the proof, it remains to use Theorem 1.9.7.   □

As a corollary, we obtain the following theorem.

Theorem 1.9.10 (Abel) *A generic algebraic equation of degree 5 or greater is not solvable by radicals.*

Remark 1.9.11 Abel had proved his theorem by a different method even before Galois theory appeared. His approach was later developed by Liouville. Liouville's method allows, for example, to prove that many elementary integrals cannot be computed by elementary functions.

Arnold proved topologically that a general algebraic equation of degree greater than 4 over the field of rational functions of one complex variable is not solvable by radicals [Alekseev 04]. I constructed a topological variant of Galois theory that allows one to prove that a general algebraic equation of degree $k > 4$ over the field of rational functions of several complex variables cannot be solved using all elementary and meromorphic functions of several variables, composition, arithmetic operations, integration, and solutions of algebraic equations of degree less than k.

1.10 Unsolvability of Complicated Equations by Solving Simpler Equations

Is it possible to solve a given complicated algebraic equation using the solutions of other, simpler, algebraic equations as admissible operations? We have considered two well-posed questions of this kind: the question of solvability of equations by radicals (in which the simpler equations are those of the form $x^n - a = 0$) and the question of solvability of equations by k-radicals (in which the simpler equations are those of the form $x^n - a = 0$ and all algebraic equations of degree k or less). In this section, we consider the general question of solvability of complicated equation by solving simpler equations. In Sect. 1.10.1, we set up the problem of B-solvability of equations and discuss a necessary condition for the solvability. In Sect. 1.10.2, we discuss classes of groups connected to the problem of B-solvability of equations.

1.10.1 A Necessary Condition for Solvability

Let B be a collection of algebraic equations. An algebraic equation defined over a field K is automatically defined over any bigger field K_1, $K \subset K_1$. We will assume that the collection B of algebraic equations, together with any equation defined over a field K, contains the same equation considered as an equation over any bigger field $K_1 \supset K$.

Definition 1.10.1 An algebraic equation over a field K is said to be *solvable by solving equations from the collection B*, or *B-solvable* for short, if there exists a chain of fields $K = K_0 \subset K_1 \subset \cdots \subset K_n$ such that all roots of the equation belong to the field K_n, and for every $i = 0, \ldots, n - 1$, the field K_{i+1} is obtained from the field K_i by adjoining all roots of some algebraic equation from the collection B defined over the field K_i.

Is a given algebraic equation B-solvable? Galois theory provides a necessary condition for B-solvability of equations. In this subsection, we discuss this condition. To the collection B of equations, we assign the set $G(B)$ of Galois groups of these equations.

Proposition 1.10.2 *The set $G(B)$ contains, together with any finite group, all subgroups of it.*

Proof Suppose that some equation defined over the field K belongs to the collection B. Let P be the field obtained from K by adjoining all roots of this equation, G the Galois group of the field P over the field K, and $G_1 \subset G$ any subgroup. Let K_1 denote the intermediate field corresponding to the subgroup G_1. The Galois group of our equation over the field K_1 coincides with G_1. By our assumption, the collection B, together with any equation defined over the field K, contains the same equation defined over the bigger field K_1.   □

Theorem 1.10.3 (A necessary condition for B-solvability) *If an algebraic equation over a field K is B-solvable, then its Galois group G admits a normal tower $G = G_0 \supset G_1 \cdots \supset G_1 = \{e\}$ of subgroups in which every quotient G_i/G_{i+1} is a quotient of some group from $B(G)$.*

Proof Indeed, the B-solvability of an equation over the field K means the existence of a chain of extensions $K = K_0 \subset K_1 \subset \cdots \subset K_n$ in which the field K_{i+1} is obtained from the field K_i by adjoining all roots of some equation from B, and the last field K_n contains all roots of the initial algebraic equation. Let $G = G_0 \supset \cdots \supset G_n = \{e\}$ be the chain of Galois groups of this equation over this chain of subfields. We will show that the chain of subgroups thus obtained satisfies the requirements of the theorem. Indeed, by Theorem 1.7.10, the group G_{i+1} is a normal subgroup of the group G_i; moreover, the quotient group G_i/G_{i+1} is simultaneously a quotient of the Galois group of the field K_{i+1} over the field K_i. Since the field K_{i+1} is obtained from the field K_i by adjoining all roots of some equation from B, the Galois group of the field K_{i+1} over the field K_i belongs to the set $G(B)$. $\square$

1.10.2 Classes of Finite Groups

Let M be a set of finite groups.

Definition 1.10.4 Define the *completion* $\mathscr{K}(M)$ of the set M as the minimal class of finite groups containing all groups from M and satisfying the following properties:

1. together with any group, the class $\mathscr{K}(M)$ contains all subgroups of it;
2. together with any group, the class $\mathscr{K}(M)$ contains all quotients of it;
3. if a group G has a normal subgroup H such that the groups H and G/H are in the class $\mathscr{K}(M)$, then the group G is in the class $\mathscr{K}(M)$.

The theorem proved above suggests the following problem: for a given set M of finite groups, describe its completion $\mathscr{K}(M)$. Recall the Jordan–Hölder theorem. A normal tower $G = G_0 \supset \cdots \supset G_n = \{e\}$ of a group G is said to be *unrefinable* (or *maximal*) if all quotient groups G_i/G_{i+1} with respect to this tower are simple groups. The Jordan–Hölder theorem asserts that *for every finite group G, the set of quotient groups with respect to any unrefinable normal tower of the group G does not depend on the choice of an unrefinable tower* (and hence is an invariant of the group).

Proposition 1.10.5 *A group G belongs to the class $\mathscr{K}(M)$ if and only if every quotient group G_i/G_{i+1} with respect to an unrefinable normal tower of the group G is a quotient group of a subgroup of a group from M.*

Proof Firstly, by definition of the class $\mathscr{K}(M)$, every group G satisfying the assumptions of the proposition belongs to the class $\mathscr{K}(M)$. Secondly, it is not hard to verify that groups G satisfying the assumptions of the proposition have properties 1–3 from the definition of the completion of M. $\qquad\square$

Corollary 1.10.6

1. *The completion of the class of all finite Abelian groups is the class of all finite solvable groups.*
2. *The completion of the set of groups consisting of all Abelian groups and the group $S(k)$ is the class of all finite k-solvable groups.*

Remark 1.10.7 Necessary conditions for solvability of algebraic equations by radicals and by k-radicals are particular cases of Theorem 1.10.3.

1.11 Finite Fields

While proving some theorems of Galois theory we assumed that the field of coefficients is infinite (this was essential in Proposition 1.2.1, which was used in many constructions). Here we establish Galois theory for finite fields (for such fields it has an especially simple and complete form). This subsection is practically independent of the others and uses only properties of the group of roots of unity that lie in a given field (Proposition 1.8.2) and algebraicity of elements over the field of invariants under the action of a finite automorphism group (Theorem 1.3.3, part 2).

Every field P is a vector space over a subfield K of P. Every finite field P is a finite-dimensional vector space over a subfield K.

Lemma 1.11.1 *Let k be the dimension of a finite field P over a subfield K, and q the number of elements in K. Then the field P contains q^k elements.*

Proof Every element a of the field P can be uniquely represented in the form $\lambda_1 e_1 + \cdots + \lambda_k e_k$, where $e_1, \ldots, e_k$ is a basis in P over K, and $\lambda_1, \ldots, \lambda_k \in K$. $\qquad\square$

Corollary 1.11.2

1. *Let the characteristic of a finite field P be equal to p. Then the field P has p^k elements, where k is a positive integer.*
2. *Let K be a subfield of the field P having p^k elements. Then the number of elements in the field K is equal to p^m, where m is a divisor of the number k, $k = ml$, $l \in \mathbb{Z}$.*

Proof The additive subgroup of the field P spanned by the unit element is a subfield isomorphic to the field $\mathbb{Z}/p\mathbb{Z}$, where p is the characteristic of the field P. By Lemma 1.11.1 the field P contains p^k elements. In the same way, the subfield $K \subset P$ contains p^m elements (here k and m are dimensions of the spaces

P and K over the field $\mathbb{Z}/p\mathbb{Z}$). By Lemma 1.11.1, the number p^k is a power of p^m. Hence, m is a divisor of k. $\square$

Lemma 1.11.3 *Let a finite field P contain q elements. Then every nonzero element a of the field P satisfies the equation $a^{q-1} = 1$. For every element a of the field P the identity $a^q = a$ holds. The multiplicative group P^* of the field P is a cyclic group of order $(q - 1)$.*

Proof The multiplicative group $P^* \subset P$ of nonzero elements of the field P has order $q - 1$. The order of the group is divisible by the order of any of its elements. This implies the relation $a^{q-1} = 1$. Multiplying this relation by a, we get the identity $a^q = a$, which also holds for the zero element. Every finite subgroup of roots of unity in any field is a cyclic subgroup (Proposition 1.8.2). Therefore, the group P^* is cyclic. $\square$

The *Frobenius homomorphism* $F : P \to P$ of the field P of characteristic p is the homomorphism such that $F(a) = a^p$. *For finite fields, the Frobenius homomorphism is an automorphism.* Indeed, $F(a) = 0$ if and only if $a = 0$, and a map F of a finite set into itself such that $F(a) \neq F(b)$ for $a \neq b$ is bijective.

Theorem 1.11.4 *For a prime number p and a positive integer k there exists a unique finite field P containing p^k elements. The Frobenius automorphism generates the cyclic group $G = \{F, F^2, \ldots, F^k = \mathrm{Id}\}$ of automorphisms of the field P. For every divisor m of the number k:*

1. *there exists a unique subgroup $G_m \subset G$ of order k/m generated by the automorphism F^m;*
2. *there exists a unique subfield $K_m \subset P$ containing p^m elements and consisting of the invariants of the group G_m.*

There are no other subfields of the field P and no other subgroups of the group G.

Proof As we have shown above, the field containing p^k elements must contain the field $\mathbb{Z}/p\mathbb{Z}$ and all roots of the equation $x^{p^k} - x = 0$ over this field. Consider the field $\tilde{P}$ obtained by adjoining all roots of the equation $x^{p^k} - x = 0$ to the field $\mathbb{Z}/p\mathbb{Z}$ (i.e., $\tilde{P}$ is the splitting field of the polynomial $x^{p^k} - x$ over the field $\mathbb{Z}/p\mathbb{Z}$). The field $\tilde{P}$ has characteristic p. The set of elements of the field that are invariant under the action of the kth power of the Frobenius automorphism F is a field. The elements of the field invariant under the action of F^k are roots of the equation $x^{p^k} = x$. Hence, all roots of the equation $x^{p^k} = x$ over the field $\mathbb{Z}/p\mathbb{Z}$ form a subfield of $\tilde{P}$. By definition of $\tilde{P}$, this field coincides with the field $\tilde{P}$. It follows that there exists a unique field of p^k elements; this is the field $\tilde{P}$ (i.e., the splitting field of the polynomial $x^{p^k} - x$ over the field $\mathbb{Z}/p\mathbb{Z}$).

By Corollary 1.11.2, a subfield K of the field P must contain $q = p^n$ elements, where $p^k = q^l$ and l is a positive integer. Let us show that there exists a unique subfield with this number of elements. Let a denote a generator of the group P^*. The generator a of the group P^* has order $q^l - 1$. The number $q^l - 1$ is divisible by $q - 1$. Indeed, $q^l - 1 = (q - 1)n$, where $n = q^l + \cdots + 1$. Put $b = a^n$. The element b has order $q - 1$. The elements $1, b, \ldots, b^{q-2}$, and 0 are the roots of the equation $x^q = x$, i.e., they form the field of invariants of the automorphism F^m (and of the group of automorphisms generated by F^m). These elements form a subfield $K \subset P$ containing q elements. Conversely, the elements of any subfield containing q elements must be precisely all roots of the equation $x^q = x$. Theorem 1.11.4 is proved. $\square$

Let us list all automorphisms of a finite field P.

Theorem 1.11.5 *Under the hypothesis of Theorem* 1.11.4, *the group G coincides with the group of all automorphisms of the field P.*

Proof Every automorphism of the field P fixes 0 and 1. Hence, every automorphism of the field P fixes the subfield $\mathbb{Z}/p\mathbb{Z}$ pointwise. Consider the field of invariants with respect to the action of the group G. It consists of all roots of the equation $x^p = x$ and thus coincides with the field $\mathbb{Z}/p\mathbb{Z}$. Let a be a generator of the cyclic group P^*. Since the group G contains k elements, the generator (as well as any other element of the field P) satisfies a polynomial equation of degree at most k over the field $\mathbb{Z}/p\mathbb{Z}$ (Theorem 1.3.3, part 2). The image of a under the action of an automorphism defines uniquely that automorphism, since every other element of P^* is a power of a. Hence there exist at most k distinct automorphisms of the field P. We know k automorphisms of the field P; these are the powers of the automorphism F. It follows that the field P does not have any other automorphisms. $\square$

Corollary 1.11.6 *Theorem* 1.11.4 *establishes a one-to-one Galois correspondence between the subgroups G_m of the Galois group G of the field P over the subfield $\mathbb{Z}/p\mathbb{Z}$ and the intermediate subfields K_m.*

Let a field K have $q = p^l$ elements, and let a field P have q^k elements and contain the field K. Let G_l denote the group of automorphisms of the field P generated by the lth power of the Frobenius automorphism F.

Corollary 1.11.7 *The field K is the field of invariants for the group G_l. Subfields of the field P containing the field K are in one-to-one correspondence with the subgroups of the group G_l.*

Proof Every subgroup of the group G_l is a cyclic group generated by an element $(F^l)^m$, where m is a divisor of the number k. This subgroup fixes pointwise the intermediate field containing q^m elements. $\square$

Let K be a finite field containing $q = p^l$ elements, $Q \in K[x]$ an irreducible polynomial of degree k over the field K, and P the splitting field of Q containing q^k elements.

Corollary 1.11.8 *All roots* $y_1, \ldots, y_k$ *of the equation* $Q = 0$ *in the field* P *are simple roots. They can be labeled so that the equality* $y_{i+1} = y_i^{p^l}$ *holds if* $1 \le i < k$ *and* $y_k^{p^l} = y_1$. *The equation* $Q = 0$ *is separable over the field* K *and is a Galois equation over this field.*

Proof The cyclic group G_l acts on the field P with the field of invariants K. The orbit of the root y_1 under the action of this group consists of the elements $y_1, \ldots, y_k$ such that $y_{i+1} = y_i^{p^l}$ holds if $1 \le i < k$ and $y_k^{p^l} = y_1$. According to the Theorem 1.3.3, part 2, the roots of the polynomial Q are simple and coincide with the orbit $y_1, \ldots, y_k$. Every root of the equation $Q = 0$ is a power of every other root of this equation. Hence, the equation $Q = 0$ is a Galois equation.   □

Chapter 2
Coverings

This chapter is devoted to coverings. There is a surprising analogy between the classification of coverings over a connected, locally connected, and simply connected topological space and Galois theory. We state the classification results for coverings so that their formal similarity with Galois theory becomes evident.

There is a whole series of closely related problems on classification of coverings. Apart from the usual classification, there is a classification of coverings with marked points. One can fix a normal covering and classify coverings (and coverings with marked points) that are subordinate to this normal covering. For our purposes, it is necessary to consider ramified coverings over Riemann surfaces and to solve analogous classification problems for ramified coverings, etc.

In Sect. 2.1, we consider coverings over topological spaces. We discuss in detail the classification of coverings with marked points over a connected, locally connected, and simply connected topological space. Other classification problems reduce easily to this classification.

In Sect. 2.2, we consider finite ramified coverings over Riemann surfaces. Ramified coverings are first defined as those proper maps of real manifolds to a Riemann surface whose singularities are similar to the singularities of complex analytic maps. Then we show that ramified coverings have a natural complex analytic structure.

We discuss the operation of completion for coverings over a Riemann surface X with a removed discrete set O. This operation can be applied equally well to coverings and to coverings with marked points. It transforms a finite covering over $X \setminus O$ to a finite ramified covering over X.

Classification of finite ramified coverings with a fixed ramification set almost repeats the analogous classification of unramified coverings. Therefore, we allow ourselves to formulate results without proofs.

To compare the main theorem of Galois theory and the classification of ramified coverings, we use the following fact. The set of orbits under a finite group action on a one-dimensional complex analytic manifold has a natural structure of a complex analytic manifold. The proof uses the Lagrange resolvent (in Galois theory, the Lagrange resolvents are used to prove solvability by radicals of equations with a solvable Galois group).

A. Khovanskii, *Galois Theory, Coverings, and Riemann Surfaces*,
DOI 10.1007/978-3-642-38841-5_2, © Springer-Verlag Berlin Heidelberg 2013

At the end of this second chapter, we apply the operation of completion of coverings to define the Riemann surface of an irreducible algebraic equation over the field $K(X)$ of meromorphic functions over a manifold X.

2.1 Coverings over Topological Spaces

This section is devoted to coverings over a connected, locally connected, and simply connected topological space. There is a series of closely related problems on classification of coverings. We discuss in detail the classification of coverings with marked points. Other classification problems reduce easily to this classification.

In Sect. 2.1.1, we recall the covering homotopy theorem. In Sect. 2.1.2, we prove a classification theorem for coverings with marked points. In Sect. 2.1.3, we discuss the correspondence between subgroups of the fundamental group and coverings with marked points. In Sect. 2.1.4, we discuss other classifications of coverings and their formal similarity with Galois theory.

This section does not rely on the other parts of the book and can be read independently.

2.1.1 Coverings and Covering Homotopy

Continuous maps f_1 and f_2 from topological spaces Y_1 and Y_2, respectively, to a topological space X are called *left equivalent* if there exists a homeomorphism $h : Y_1 \to Y_2$ such that $f_1 = f_2 \circ h$. A topological space Y together with a projection $f : Y \to X$ to a topological space X is called a *covering with the fiber D* over X (where D is a discrete set) if the following holds: for each point $c \in X$ there exists an open neighborhood U such that the projection map of $U \times D$ onto the first factor is left equivalent to the map $f : Y_U \to U$, where $Y_U = f^{-1}(U)$. The following theorem holds for coverings.

Theorem 2.1.1 (Covering homotopy theorem) *Let $f : Y \to X$ be a covering, W_k a k-dimensional cellular complex, and $F : W_k \to X$, $\tilde{F} : W_k \to Y$ its mappings to X and Y such that $f \circ \tilde{F} = F$. Then for every homotopy $F_t : W_k \times [0, 1] \to X$ of the map F, $F_0 = F$, there exists a unique lifting homotopy $\tilde{F} : W_k \times [0, 1] \to Y$, $f(\tilde{F}_t) = F_t$, of the map $\tilde{F}$, $\tilde{F}_0 = F$.*

We will use this theorem in the cases in which the complex W_k is a point or the interval $[0, 1]$. Recall the proof of the covering homotopy theorem in the first case. The proof in the second case is analogous, and we will omit it. Let us formulate the first case separately.

Lemma 2.1.2 *For each curve $\gamma : [0, 1] \to X$, $\gamma(0) = a$ and for each point $b \in Y$ that is projected to a, $f(b) = a$, there exists a unique curve $\tilde{\gamma} : [0, 1] \to Y$ such that $\tilde{\gamma}(0) = b$ and $f \circ \tilde{\gamma} = \gamma$.*

Proof If Y is the direct product $Y = X \times D$, then the lemma is obvious. Consider the curve $\gamma : [0, 1] \to X$. Say that an interval of the segment $[0, 1]$ is sufficiently small with respect to γ if its image under the map γ lies in such a neighborhood in X that the cover over it is a direct product. Since the curve is compact, there exists a subdivision of the segment $[0, 1]$ into smaller segments (with common endpoints) that are sufficiently small with respect to γ. Lift to Y the piece of the curve over the first of these segments $[0, a_1]$, that is, construct the curve $\tilde{\gamma} : [0, a_1] \to Y$, $\tilde{\gamma}(0) = b$, $f \circ \tilde{\gamma} = \gamma$. Then lift to Y the curve over the second segment $[a_1, a_2]$ using the already constructed point $b_1 = \tilde{\gamma}(a_1)$. Continuing this process, we lift to Y the whole curve γ. The uniqueness of a lift of the curve is evident: First of all, the set of points in the segment $[0, 1]$ where two lifts coincide is nonempty, since by hypothesis, each lift starts at the point $b \in Y$. Second, it is open, since locally, Y is a direct product of an open subset of X by a discrete set. Third, it is closed, since curves are continuous. Hence, two lifts coincide on the whole segment $[0, 1]$. □

We now define normal coverings and groups of deck transformations, which play a central role in this chapter. Consider a covering $f : Y \to X$. A homeomorphism $h : Y \to Y$ is called a *deck transformation* of this covering if the equality $f = f \circ h$ is satisfied. Deck transformations form a group. A covering is called *normal* if its group of deck transformations acts transitively on each fiber $f^{-1}(a)$, $a \in X$, of the covering, and the following topological conditions on the spaces X and Y are satisfied: the space Y is connected, and the space X is locally connected and locally simply connected.

2.1.2 Classification of Coverings with Marked Points

A triple $f : (Y, b) \to (X, a)$ consisting of spaces with marked points (X, a), (Y, b) and a map f is called a *covering with marked points* if $f : Y \to X$ is a covering and $f(b) = a$. Coverings with marked points are equivalent if there exists a homeomorphism between covering spaces that commutes with projections and maps the marked point to the marked point. It is usually clear from the notation whether we mean coverings or coverings with marked points. In such cases, we will for brevity omit the words "with marked points" when talking about coverings.

A covering with marked points $f : (Y, b) \to (X, a)$ defines the homomorphism $f_* : \pi_1(Y, b) \to \pi_1(X, a)$ of the fundamental group $\pi_1(Y, b)$ of the space Y with the marked point b to the fundamental group $\pi_1(X, a)$ of the space X with the marked point a.

Lemma 2.1.3 *For a covering with marked points, the induced homomorphism of the fundamental groups has trivial kernel.*

Proof Let a closed path $\gamma : [0, 1] \to X$, $\gamma(0) = \gamma(1) = a$, in the space X be the image $f \circ \tilde{\gamma}$ of the closed path $\tilde{\gamma} : [0, 1] \to Y$, $\tilde{\gamma}(0) = \tilde{\gamma}(1) = b$, in the space Y. Let the path γ be homotopic to the identity path in the space of paths in X with fixed endpoints. Then the path $\tilde{\gamma}$ is homotopic to the identity path in the space of paths in Y with fixed endpoints. For the proof, it is enough to lift the homotopy with fixed endpoints to Y.   □

The following theorem holds for every connected, locally connected, and locally simply connected topological space X with a marked point a.

Theorem 2.1.4 (On classification of coverings with marked points)

1. *For every subgroup G of the fundamental group of the space X there exist a connected space (Y, b) and a covering over (X, a) by the covering space (Y, b) such that the image of the fundamental group of the space (Y, b) coincides with the subgroup G.*
2. *Two coverings over (X, a) by connected covering spaces (Y, b_1) and (Y, b_2) are equivalent if the images of the fundamental group of these spaces in the fundamental group of (X, a) coincide.*

Proof

1. Consider the space $\hat{\Omega}(X, a)$ of the paths $\gamma : [0, 1] \to X$ in X that originate at the point a, $\gamma(0) = a$, and its subspace $\hat{\Omega}(X, a, a_1)$ consisting of paths that terminate at a point a_1. On the spaces $\hat{\Omega}(X, a)$, $\hat{\Omega}(X, a, a_1)$, consider the topology of uniform convergence and the following equivalence relation. Say that paths γ_1 and γ_2 are equivalent if they terminate at the same point a_1 and if the path γ_1 is homotopic to the path γ_2 in the space $\hat{\Omega}(X, a, a_1)$ of paths with fixed endpoints. Denote by $\Omega(X, a)$ and $\Omega(X, a, a_1)$ the quotient spaces of $\hat{\Omega}(X, a)$ and $\hat{\Omega}(X, a, a_1)$ by this equivalence relation. The fundamental group $\pi_1(X, a)$ acts on the space $\Omega(X, a)$ by right multiplication (composition). For a fixed subgroup $G \subset \pi_1(X, a)$, denote by $\Omega_G(X, a)$ the space of orbits under the action of G on $\Omega(X, a)$. Points in $\Omega_G(X, a)$ are elements of the space $\hat{\Omega}(X, a, a_1)$ defined up to homotopy with fixed endpoints and up to right multiplication by elements of the subgroup G. There is a marked point $\tilde{a}$ in this space, namely, the equivalence class of the constant path $\gamma(t) \equiv a$. The map $f : (\Omega_G(X, a), \tilde{a}) \to (X, a)$ that assigns to each path its right endpoint has the required properties. We omit a proof of this fact. Note, however, that the assumptions on the space X are necessary for the theorem to be true: if X is disconnected, then the map f has no preimages over the connected components of X disjoint from the point a, and if X is not locally connected and locally simply connected, then the map $f : (\Omega_G(X, a), \tilde{a}) \to (X, a)$ may not be a local homeomorphism.
2. We now show that a covering $f : (Y, b) \to (X, a)$ such that $f_*\pi_1(Y, b) = G \subset \pi_1(X, a)$ is left equivalent to the covering constructed using the subgroup G in the first part of the proof. To a point $y \in Y$, assign any element from the space of paths $\Omega(Y, b, y)$ in Y that originate at the point b and terminate at the point y, which are defined up to homotopy with fixed endpoints. Let $\tilde{\gamma}_1$, $\tilde{\gamma}_2$ be two paths from the space $\Omega(Y, b, y)$ and let $\tilde{\gamma} = (\tilde{\gamma}_1)^{-1} \circ \tilde{\gamma}_2$ denote the path

composed from the path $\tilde{\gamma}_2$ and the path $\tilde{\gamma}_1$ traversed in the opposite direction. The path $\tilde{\gamma}$ originates and terminates at the point b; hence the path $f \circ \tilde{\gamma}$ lies in the group G. It follows that the image $f \circ \tilde{\gamma}$ of an arbitrary path $\tilde{\gamma}$ in the space $\Omega(Y, b, y)$ under the projection f is the same point of the space $\Omega_G(X, a)$ (that is, the same path in the space $\hat{\Omega}_G(X, a)$ up to homotopy with fixed endpoints and up to right multiplication by elements of the group G). In this way, we have assigned to each point $y \in Y$ a point of the space $\Omega_G(X, a)$. It is easy to check that this correspondence defines the left equivalence between the covering $f :$ $(Y, b) \to (X, a)$ and the standard covering constructed using the subgroup $G = f_* \pi_1(Y, b)$. $\qquad \square$

2.1.3 Coverings with Marked Points and Subgroups of the Fundamental Group

Theorem 2.1.4 shows that coverings with marked points over a space X with a marked point a considered up to left equivalence are classified by subgroups G of the fundamental group $\pi_1(X, a)$. Let us discuss the correspondence between coverings with marked points and subgroups of the fundamental group.

Let $f : (Y, b) \to (X, a)$ be a covering that corresponds to the subgroup $G \subset \pi_1(X, a)$, and let $F = f^{-1}(a)$ denote the fiber over the point a. We have the following lemma.

Lemma 2.1.5 *The fiber F is in bijective correspondence with the right cosets of the group $\pi_1(X, a)$ modulo the subgroup G. If a right coset h corresponds to a point c of the fiber F, then the group hGh^{-1} corresponds to the covering $f : (Y, c) \to (X, a)$ with marked point c.*

Proof The group G acts by right multiplication on the space $\Omega(X, a, a)$ of closed paths that originate and terminate at the point a and are defined up to homotopy with fixed endpoints. According to the description of the covering corresponding to the group G (see the first part of the proof of Theorem 2.1.4), the preimages of the point a with respect to this covering are orbits of the action of the group G on the space $\Omega(X, a, a)$, i.e., the right cosets of the group $\pi_1(X, a)$ modulo the subgroup G.

Let $h : [0, 1] \to X$, $h(0) = a$, be a loop in the space X, and $\tilde{h} : [0, 1] \to Y$, $f\tilde{h} = h$, the lift of this loop to Y that originates at the point b, $\tilde{h}(0) = b$, and terminates at the point c, $\tilde{h}(1) = c$. Let $G_1 \subset \pi_1(X, a)$ be the subgroup consisting of paths whose lifts to Y starting at the point c terminate at the same point c. It is easy to verify the inclusions $hGh^{-1} \subseteq G_1$, $h^{-1}G_1h \subseteq G$, which imply that $G_1 = hGh^{-1}$. $\qquad \square$

Let us say that a covering $f_2 : (Y_2, b_2) \to (X, a)$ is *subordinate* to the covering $f_1 : (Y, b_1) \to (X, a)$ if there exists a continuous map $h : (Y_1, b_1) \to (Y_2, b_2)$ compatible with the projections f_1 and f_2, i.e., such that $f_1 = f_2 \circ h$.

Lemma 2.1.6 *The covering corresponding to a subgroup G_2 is subordinate to the covering corresponding to a subgroup G_1 if and only if the inclusion $G_2 \supseteq G_1$ holds.*

Proof Suppose that $\pi_1(X, a) \supseteq G_2 \supseteq G_1$, and let $f_2 : (Y_2, b_2) \to X$ be the covering corresponding to the subgroup G_2 of $\pi_1(X, a)$. By Lemma 2.1.3, the group G_2 coincides with the image $f_{2*}\pi_1(Y_2, b_2)$ of the fundamental group of the space Y_2 in $\pi_1(X, a)$. Let $g : (Y_1, b_1) \to (Y_2, b_2)$ be the covering corresponding to the subgroup $f_{2*}^{-1}G_1$ in the fundamental group $\pi_1(Y_2, b_2) = f_{2*}^{-1}G_2$. The map $f_2 \circ g : (Y_1, b_1) \to (X, a)$ defines the covering over (X, a) corresponding to the subgroup $G_1 \subset \pi_1(X, a)$. Hence, the covering $f_2 \circ g : (Y_2, b_2) \to (X, a)$ is left equivalent to the covering $f_1 : (Y_1, b_1) \to (X, a)$. We have proved the lemma in one direction. The proof in the opposite direction is similar. $\qquad\square$

Consider a covering $f : Y \to X$ such that Y is connected, and X is locally connected and locally simply connected. Suppose that for a point $a \in X$, the covering has the following properties: for all choices of preimages b and c of the point a, the coverings with marked points $f : (Y, b) \to (X, a)$ and $f : (Y, c) \to (X, a)$ are equivalent. Then

1. the covering has this property for every point $a \in X$;
2. the covering $f : Y \to X$ is normal.

Conversely, if the covering is normal, then it has this property for every point $a \in X$. This statement follows immediately from the definition of a normal covering.

Lemma 2.1.7 *A covering is normal if and only if it corresponds to a normal subgroup H of the fundamental group $\pi_1(X, a)$. For this normal subgroup, the group of deck transformations is isomorphic to the quotient group $\pi_1(X, a)/H$.*

Proof Suppose that the covering $f : (Y, b) \to (X, a)$ corresponding to a subgroup $G \subset \pi_1(X, a)$ is normal. Then for every preimage c of the point a, this covering is left equivalent to the covering $f : (Y, c) \to (X, a)$. By Lemma 2.1.5, this means that the subgroup G coincides with each of its conjugate subgroups. It follows that the group G is a normal subgroup of the fundamental group. Similarly, one can show that if G is a normal subgroup of the fundamental group, then the covering corresponding to this subgroup is normal.

A deck homeomorphism that takes the point b to the point c is unique. Indeed, the set on which two such homeomorphisms coincide is, firstly, open (since f is a local homeomorphism), and secondly, closed (since homeomorphisms are continuous), and thirdly, nonempty (since it contains the point b). Since the space Y is connected, this set must coincide with Y.

The fundamental group $\pi_1(X, a)$ acts by right multiplication on the space $\Omega(X, a)$. For every normal subgroup H, this action gives rise to an action on the equivalence classes in $\Omega_H(X, a)$ (under multiplication by an element $g \in \pi_1(X, a)$, the equivalence class xH is mapped to the equivalence class $xHg = xgH$). The action of the fundamental group on $\Omega_H(X, a)$ is compatible with the projection $f : \Omega_H(X, a) \to (X, a)$ that assigns to each path the point at which it terminates. Hence, the fundamental group $\pi_1(X, a)$ acts on the space Y of the normal covering $f : (Y, b) \to (X, a)$ by deck homeomorphisms. For the covering that corresponds to the normal subgroup H, the kernel of this action is the group H, i.e., there is an effective action of the quotient group $\pi_1(X, a)/H$ on the space of this covering. The quotient group action can map the point b to any other preimage c of the point a. Hence, there are no other deck homeomorphisms $h : Y \to Y$ apart from the homeomorphisms of the action of the quotient group $\pi_1(X, a)/H$. The lemma is proved. $\square$

The fundamental group $\pi_1(X, a)$ acts on the fiber $F = f^{-1}(a)$ of the covering $f : (Y, b) \to (X, a)$. We now define this action. Let γ be a path in the space X that originates and terminates at the point a. For every point $c \in F$, let $\tilde{\gamma}_c$ denote a lift of the path γ to Y such that $\tilde{\gamma}_c(0) = c$. The map $S_\gamma : F \to F$ that takes the point c to the point $\tilde{\gamma}_c(1) \in F$ belongs to the group $S(F)$ of bijections from the set F to itself. The map S_γ depends only on the homotopy class of the path γ, that is, on the element of the fundamental group $\pi_1(X, a)$ represented by the path γ. The homomorphism $S_\gamma : \pi_1(X, a) \to S(F)$ is called the *monodromy homomorphism*, and the image of the fundamental group in the group $S(F)$ is called the *monodromy group* of the covering $f : (Y, b) \to (X, a)$.

Let $f : (Y, b) \to (X, a)$ be the covering corresponding to a subgroup $G \subset \pi_1(X, a)$, $F = f^{-1}(a)$ the fiber of this covering over the point a, and $S(F)$ the permutation group of the fiber F. We have the following lemma.

Lemma 2.1.8 *The monodromy group of the covering described in the previous paragraph is a transitive subgroup of the group $S(F)$ and is equal to the quotient group of $\pi_1(X, a)$ by the largest normal subgroup H that is contained in the group G, i.e.,*

$$H = \bigcap_{h \in \pi_1(X,a)} hGh^{-1}.$$

Proof The monodromy group is transitive. For the proof, we have to construct, for every point $c \in F$, a path γ such that $S_\gamma(b) = c$. Take an arbitrary path $\tilde{\gamma}$ in the connected space Y such that $\tilde{\gamma}$ connects the point b with the point c. To obtain the path γ it is enough to take the image of the path $\tilde{\gamma}$ under the projection f.

It can be immediately seen from the definitions that the stabilizer of the point b under the action of the fundamental group on the fiber F coincides with the group $G \subset \pi_1(X, a)$. Let $h \in \pi_1(X, a)$ be an element in the fundamental group that takes the point b to the point $c \in F$. Then the stabilizer of the point c is equal to hGh^{-1}. The kernel H of the monodromy homomorphism is the intersection of the stabilizers of all points in the fiber, that is, $H = \bigcap_{h \in \pi_1(X,a)} hGh^{-1}$. The intersection of all groups hGh^{-1} is the largest normal subgroup contained in the group G. $\square$

2.1.4 Coverings and Galois Theory

In this subsection, we discuss an analogy between coverings and Galois theory. First, we discuss the usual classification of coverings (without marked points). Then we prove the classification theorem for coverings and coverings with marked points subordinate to a given normal covering. This theorem is surprisingly similar to the main theorem of Galois theory. To make this analogy more transparent we introduce the notion of intermediate coverings and reformulate the classification theorem for the intermediate coverings. At the end of the subsection, we give one more description of intermediate coverings that directly relates such coverings to the subgroups of the deck transformation group acting on the normal covering.

We now pass to coverings without marked points. We classify coverings with the connected covering space over a connected, locally connected, and simply connected space. This classification reduces to the analogous classification of coverings with marked points.

Two coverings $f_1 : Y_1 \to X$ and $f_2 : Y_2 \to X$ are called equivalent if there exists a homeomorphism $h : Y_1 \to Y_2$ compatible with the projections f_1 and f_2, i.e., such that $f_1 = f_2 \circ h$.

Lemma 2.1.9 *Coverings with marked points are equivalent as coverings* (rather than as coverings with marked points!) *if and only if the subgroups corresponding to these coverings are conjugate in the fundamental group of the space.*

Proof Let coverings $f_1 : (Y_1, b_1) \to (X, a)$ and $f_2 : (Y_2, b_2) \to (X, a)$ be equivalent as coverings. A homeomorphism h should map the fiber $f_1^{-1}(a)$ to the fiber $f_2^{-1}(a)$. Hence, the covering $f_1 : (Y_1, b_1) \to (X, a)$ is equivalent, as a covering with marked points, to the covering $f_2 : (Y_2, h(b_1)) \to (X, a)$, where $f_2(h(b_1)) = f_2(b_2)$. This means that the subgroups corresponding to the original coverings with marked points are conjugate. $\square$

Therefore, *coverings* $f : Y \to X$ such that Y is connected and X is locally connected and locally simply connected *are classified by subgroups of the fundamental group* $\pi_1(X)$ *defined up to conjugation in the group* $\pi_1(X)$ (the groups $\pi_1(X, a)$ obtained by different choices of the base point are isomorphic, and isomorphism is well defined up to conjugation).

In Galois theory, one usually considers algebraic extensions that belong to a given Galois extension (and not all field extensions simultaneously). Analogously, when classifying coverings and coverings with marked points, one can restrict attention to coverings subordinate to a given normal covering.

The definition of the subordination relation for coverings with marked points was given above. One can also define an analogous relation for coverings, at least in the case that one of the coverings is normal.

Say that a *covering* $f : Y \to X$ *is subordinate to the normal covering* $g : M \to X$ if there exists a map $h : M \to Y$ compatible with the projections g and f, i.e., such that $g = f \circ h$. It is clear that a *covering is subordinate to a normal covering if and only if every subgroup from the corresponding class of conjugate subgroups in the fundamental group of X contains the normal subgroup corresponding to the normal covering.*

Fix a marked point a in the space X. Let $g : (M, b) \to (X, a)$ be the normal covering corresponding to a normal subgroup H of the group $\pi_1(X, a)$, and $N = \pi_1(X, a)/H$ the deck transformation group of this normal covering. Consider all possible coverings and coverings with fixed points subordinate to this normal covering. We can apply all classification theorems to these coverings. The role of the fundamental group $\pi_1(X, a)$ will be played by the deck transformation group N of the normal covering.

Let $f : (Y, b) \to (X, a)$ be a subordinate covering with marked points, and G the corresponding subgroup of the fundamental group. With this subordinate covering, we associate the subgroup of the deck transformation group N equal to the image of the subgroup G under the quotient projection $\pi(X, a) \to N$. The following theorem holds for this correspondence.

Theorem 2.1.10 *The correspondence between coverings with marked points subordinate to a given normal covering and subgroups of the deck transformation group of this normal covering is bijective.*

Subordinate coverings with marked points are equivalent as coverings if and only if the corresponding subgroups are conjugate in the deck transformation group.

A subordinate covering is normal if and only if it corresponds to a normal subgroup M of the deck transformation group N. The deck transformation group of the subordinate normal covering is isomorphic to the quotient group N/M.

Proof For the proof, it is enough to apply the already proved "absolute" classification results and the following evident properties of the group quotients.

The quotient projection is a bijection between all subgroups of the original group that contain the kernel of the projection and all subgroups of the quotient group. This bijection

1. preserves the partial order on the set of subgroups defined by inclusion;
2. takes a class of conjugate subgroups of the original group to a class of conjugate subgroups of the quotient group;
3. establishes a one-to-one correspondence between all normal subgroups of the original group that contain the kernel of the projection and all normal subgroups of the quotient group.

Under the correspondence of normal subgroups described in item 3, the quotient of the original group by a normal subgroup is isomorphic to the quotient of the quotient group by the corresponding normal subgroup. $\square$

Subordinate coverings are classified in two different ways: as coverings with marked points and as coverings. What are analogous classification results in Galois theory? To make the answer to this question evident let us formulate the classification problem for coverings avoiding the use of marked points.

Let $f : M \rightarrow X$ be a normal covering (as usual, we assume that the space M is connected, and the space X is locally connected and locally simply connected). An *intermediate* covering between M and X is a space Y together with a surjective and continuous map $h_Y : M \rightarrow Y$ and a projection $f_Y : Y \rightarrow X$ satisfying the condition $f = f_Y \circ h_Y$.

Let us introduce two different notions of equivalence for intermediate coverings. Say that two intermediate coverings

$$M \xrightarrow{h_1} Y_1 \xrightarrow{f_1} X \quad \text{and} \quad M \xrightarrow{h_2} Y_2 \xrightarrow{f_2} X$$

are equivalent as *subcoverings of the covering* $f : M \rightarrow X$ if there exists a homeomorphism $h : Y_1 \rightarrow Y_2$ that makes the diagram

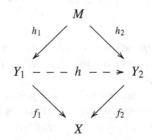

commutative, i.e., such that $h_2 = h \circ h_1$ and $f_1 = f_2 \circ h$. Say that two subcoverings are equivalent as *coverings over* X if there exists a homeomorphism $h : Y_1 \rightarrow Y_2$ such that $f_1 = h \circ f_2$ (the homeomorphism h is not required to make the upper part of the diagram commutative).

The classification of intermediate coverings regarded as subcoverings is equivalent to the classification of subordinate coverings with marked points. Indeed, if we mark a point b in the space M that lies over the point a, then we obtain a canonically defined marked point $h_Y(a)$ in the space Y.

The following statement is a reformulation of Theorem 2.1.10.

Proposition 2.1.11 *Intermediate coverings for a normal covering with the deck transformation group N:*

1. *regarded as subcoverings are classified by subgroups of the group N;*
2. *regarded as coverings over X are classified by the conjugacy classes of subgroups in the group N.*

A subordinate covering is normal if and only if it corresponds to a normal subgroup M of the deck transformation group N. The deck transformation group of the subordinate normal covering is isomorphic to the quotient group N/M.

The classification of intermediate coverings subordinate to a given normal covering is formally analogous to the classification of intermediate subfields of a given Galois extension. To see this, replace the words "normal covering," "deck transformation group," "subordinate covering" with the words "Galois extension," "Galois group," "intermediate field."

An intermediate field K_1 lying between a field K and a Galois extension P of K can be considered from two different viewpoints: as a subfield of the field P and as an extension of the field K. The classification of intermediate coverings regarded as subcoverings corresponds to Galois-theoretic classification of intermediate extensions regarded as subfields of the field P. The classification of intermediate coverings regarded as coverings over X corresponds to the Galois-theoretic classification of intermediate extensions regarded as extensions of the field P.

In Sect. 2.2 we consider finite ramified coverings over one-dimensional complex manifolds. Ramified coverings (with marked points or without marked points) over a manifold X whose ramification points lie over a given discrete set O are classified in the same way as coverings (with marked points or without marked points) over $X \setminus O$ (see Sect. 2.2). Finite ramified coverings correspond to algebraic extensions of the field of meromorphic functions on X. The fundamental theorem of Galois theory for these fields and the classification of intermediate coverings are not only formally similar but also very close to each other.

Let us give yet another description of intermediate coverings for a normal covering $f : M \to X$ with a deck transformation group N. The group N is a group of homeomorphisms of the space M with the following discreteness property: each point of the space M has a neighborhood such that its images under the action of different elements of the group N do not intersect. To construct such a neighborhood, take a connected component of the preimage under the projection $f : M \to X$ of a connected and locally connected neighborhood of the point $f(z) \in X$.

For every subgroup G of the group N, consider the quotient space M_G of the space M under the action of the group G. A point in M_G is an orbit of the action of the group G on the space M. The topology on M_G is induced by the topology on the space M. A neighborhood of an orbit consists of all orbits that lie in an invariant open subset U of the space M with the following properties. The set U contains the original orbit, and a connected component of the set U intersects each orbit at most at one point. The space M_N can be identified with the space X. To do so, we identify a point $x \in X$ with the preimage $f^{-1}(x) \subset M$, which is an orbit of the deck transformation group N acting on M. Under this identification, the quotient projection $f_{e,N} : M \to M_N$ coincides with the original covering $f : M \to X$.

Let G_1, G_2 be two subgroups in N such that $G_1 \subseteq G_2$. Define the map $f_{G_1,G_2} : M_{G_1} \to M_{G_2}$ by assigning to each orbit of the group G_1 the orbit of the group G_2 that contains it. It is easy to see that the following hold:

1. The map f_{G_1,G_2} is a covering.
2. If $G_1 \subseteq G_2 \subseteq G_3$, then $f_{G_1,G_2} = f_{G_2,G_3} f_{G_1,G_2}$.
3. Under the identification of M_N with X, the map $f_{G,N} : M_G \to M_N$ corresponds to a covering subordinate to the original covering $f_{e,N} : M \to M_N$ (since $f_{e,N} = f_{G,N} \circ f_{e,G}$).
4. If G is a normal subgroup in N, then the covering $f_{G,N} : M_G \to M_N$ is normal, and its deck transformation group is equal to N/G.

With an intermediate covering $f_{G,N} : M_G \to M_N$, one can associate either the triple of spaces $M \xrightarrow{f_{e,G}} M_G \xrightarrow{f_{G,N}} M_N$ with the maps $f_{e,G}$ and $f_{G,N}$, or the pair of spaces $M_G \xrightarrow{f_{G,N}} M_N$ with the map $f_{G,N}$. These two possibilities correspond to two viewpoints with respect to an intermediate covering, regarding it either as a subcovering or as a covering over M_N.

2.2 Completion of Finite Coverings over Punctured Riemann Surfaces

In this section, we consider finite ramified coverings over one-dimensional complex manifolds. We describe the operation of completion for coverings over a one-dimensional complex manifold X with a removed discrete set O. This operation can be applied equally well to coverings and to coverings with marked points. It transforms a finite covering over $X \setminus O$ to a finite ramified covering over X.

In Sect. 2.2.1, we consider the local case in which coverings of an open punctured disk get completed. In the local case, the operation of completion allows us to prove the Puiseux expansion for multivalued functions with an algebraic singularity.

In Sect. 2.2.2, we consider the general case. First, we define the real operation of filling holes. Then we show that the ramified covering obtained using the real operation of filling holes has a natural structure of a complex manifold.

In Sect. 2.2.3, we classify finite ramified coverings with a fixed ramification set. The classification literally repeats the analogous classification of unramified coverings. Therefore, we allow ourselves to formulate results without proofs. We prove that the set of orbits under a finite group action on a one-dimensional complex analytic manifold has a natural structure of a complex analytic manifold.

In Sect. 2.2.4, we apply the operation of completion of coverings to define the Riemann surface of an irreducible algebraic equation over the field $K(X)$ of meromorphic functions over a manifold X.

Section 2.2.2 relies on the results of Sect. 2.2.1.

2.2.1 Filling Holes and Puiseux Expansions

Let D_r be an open disk of radius r on the complex line with center at the point 0, and let $D_r^* = D_r \setminus \{0\}$ denote the punctured disk. For every positive integer k, consider the punctured disk D_q^*, where $q = r^{1/k}$, together with the map $f : D_q^* \to D_r^*$ given by the formula $f(z) = z^k$.

Lemma 2.2.1 *There exists a unique (up to left equivalence) connected k-fold covering $\pi : V^* \to D_r^*$ over the punctured disk D_r^*. This covering is normal. It is equivalent to the covering $f : D_q^* \to D_r^*$, where the map f is given by the formula $x = f(z) = z^k$.*

Proof The fundamental group of the domain D^* is isomorphic to the additive group $\mathbb{Z}$ of integers. The only subgroup in $\mathbb{Z}$ of index k is the subgroup $k\mathbb{Z}$. The subgroup $k\mathbb{Z}$ is a normal subgroup in $\mathbb{Z}$. The covering $z \to z^k$ of the punctured disk D_q^* over the punctured disk D_r^* is normal and corresponds to the subgroup $k\mathbb{Z}$. □

Let $\pi : V^* \to D_r^*$ be a connected k-fold covering over a punctured disk D_r^*. Let V denote the set consisting of the domain V and a point A. We can extend the map π to the map of the set V onto the disk D_r by putting $\pi(A) = 0$. We introduce the coarsest topology on the set V such that the following conditions are satisfied:

1. The identification of the set $V \setminus \{A\}$ with the domain V is a homeomorphism.
2. The map $\pi : V \to D$ is continuous.

Lemma 2.2.2 *The map $\pi : V \to D_r$ is left equivalent to the map $f : D_q \to D_r$ defined by the formula $x = f(z) = z^k$. In particular, V is homeomorphic to the open disk D_q.*

Proof Let $h : D_q^* \to V^*$ be the homeomorphism that establishes an equivalence of the covering $\pi : V^* \to D_r^*$ and the standard covering $f : D_q^* \to D_r^*$. Extend h to the map of the disk D_q to the set V by putting $h(0) = A$. We have to check that the extended map h is a homeomorphism. Let us check, for example, that h is a continuous map. By definition of the topology on V, every neighborhood of the point A contains a neighborhood V_0 of the form $V_0 = \pi^{-1}(U_0)$, where U_0 is a neighborhood of the point 0 on the complex line. Let $W_0 \subset D_q$ be the open set defined by the formula $W_0 = f^{-1}(U_0)$. We have $h^{-1}(V_0) = W_0$, which proves the continuity of the map h at the point 0. The continuity of the map h can be proved analogously. □

We will use the notation of the preceding lemma.

Lemma 2.2.3 *The manifold V has a unique structure of an analytic manifold such that the map $\pi : V \to D_r$ is analytic. This structure is induced from the analytic structure on the disk D_q by the homeomorphism $h : D_q \to V$.*

Proof The homeomorphism h transforms the map π into the analytic map $f(z) \to z^k$. Hence, the analytic structure on V induced by the homeomorphism satisfies the condition of the lemma. Consider another analytic structure on V. The map $h : D \to V$ outside of the point 0 can be locally represented as $h(z) = \pi^{-1}z^k$ and therefore is analytic. Thus the map $h : D \to V$ is continuous and analytic everywhere except at the point 0. By the removable singularity theorem, it is also analytic at the point 0, and therefore there is a unique analytic structure on V such that the projection π is analytic. □

The transition from the real manifold V^* to the real manifold V and the transition from the covering $\pi : V^* \to D_r^*$ to the map $\pi : V \to D_r$ will be called the *real* *operation of filling a hole*. Lemma 2.2.3 shows that after a hole has been filled, the manifold V has a unique structure of a complex analytic manifold such that the map $\pi : V \to D_r$ is analytic. The transition from the complex manifold V^* to the complex manifold V and the transition from the analytic covering $\pi : V^* \to D_r^*$ to the analytic map $\pi : V \to D_r$ will be called the *operation of filling a hole*. In what follows, we will use precisely this operation.

The operation of filling a hole is intimately related to the definition of an algebraic singular point and to Puiseux series. Let us discuss this in more detail. We say that an analytic germ φ_a at a point $a \in D$ defines a multivalued function on the disk D_r with an *algebraic singularity* at the point 0 if the following conditions are satisfied:

1. The germ φ_a can be extended along any path that originates at the point a and lies in the punctured disk D_r^*.
2. The multivalued function φ in the punctured disk D_r^* obtained by extending the germ φ_a along paths in D_r^* takes a finite number k of values.
3. When approaching the point 0, the multivalued function φ grows no faster than a power function, i.e., there exist positive real numbers C, N such that any of the values of the multivalued function φ satisfies the inequality $|\varphi(x)| < C|x|^{-N}$.

Lemma 2.2.4 *A multivalued function φ with an algebraic singularity in the punctured disk D_r^* can be represented in this disk by the Puiseux series*

$$\varphi(x) = \sum_{m > -m_0} c_m x^{m/k}.$$

Proof If the function φ can be extended analytically along all paths in the punctured disk D_r^* and has k different values, then the germ $g_b = \varphi_a \circ z_b^k$, where $b^k = a$, defines a single-valued function in the punctured disk D_q^*, where $q = r^{1/k}$. By the hypothesis, the function g grows no faster than a power function when approaching the point 0. Hence in the punctured disk D_q^*, it can be represented by the Laurent series

$$g(z) = \sum_{m > -m_0} c_m z^m.$$

Substituting $x^{1/k}$ for z in the series for the function g, we obtain the Puiseux series for the function φ.　　□

2.2.2 Analytic-Type Maps and the Real Operation of Filling Holes

In this subsection, we define the real operation of filling holes. We show that the ramified covering resulting from the real operation of filling holes has a natural complex analytic structure.

Let X be a one-dimensional complex analytic manifold, M a two-dimensional real manifold, and $\pi : M \to X$ a continuous map. We say that the map π at a point $y \in M$ has an *analytic-type singularity*[1] *of multiplicity* $k > 0$ if there exist

1. a connected punctured neighborhood $U^* \subset X$ of the point $x = \pi(y)$,
2. a connected component of the domain $\pi^{-1}(U^*)$ that is a punctured neighborhood $V^* \subset M$ of the point y

such that the triple $\pi : V^* \to U^*$ is a k-fold covering. It is natural to regard the singular point y as a *multiplicity-k* preimage of the point x: the number of preimages (counted with multiplicity) of π in the neighborhood of an analytic-type singular point of multiplicity k is constant and equal to k.

A map $f : X \to M$ is an *analytic-type map*[2] if it has an analytic-type singularity at every point. Clearly, a complex analytic map $f : M \to X$ of a complex one-dimensional manifold M to a complex manifold X is an analytic-type map (when considered as a continuous map of a real manifold M to a complex manifold X). For an analytic-type map, a point y is called *regular* if its multiplicity is equal to one, and *singular* if its multiplicity k is greater than one. The set of all regular points of an analytic-type map is open. The map considered near a regular point is a local homeomorphism. The set O of singular points of an analytic-type map is a discrete subset of M.

Proposition 2.2.5 *Let M be a two-dimensional real manifold and $f : M \to X$ an analytic-type map to a one-dimensional complex analytic manifold X. Then M has a unique structure of a complex analytic manifold such that the map f is analytic.*

Proof The map f is a local homeomorphism at the points of $M \setminus O$. This local homeomorphism to the analytic manifold X makes $M \setminus O$ into an analytic manifold. Near the points of the set O, one can define an analytic structure in the same way as near the points added by the operation of filling holes. We now prove that there are no other analytic structures such that f is analytic. Let M_1 and M_2 be two copies of the manifold M with two different analytic structures. Let O_1 and O_2 be distinguished discrete subsets of M_1 and M_2, and $h : M_1 \to M_2$ a homeomorphism identifying these two copies. It is clear from the hypothesis that the homeomorphism h is analytic everywhere except at the discrete set $O_1 \subset M_1$. By the removable singularity theorem, h is a biholomorpic map. Hence the two analytic structures on M coincide. $\square$

We now return to the operation of filling holes. Let M be a real two-dimensional manifold, and $f : M \to X$ an analytic-type map of the manifold M to a complex manifold X.

Fix a local coordinate u near a point $a \in X$, $u(a) = 0$, that gives an invertible map of a small neighborhood of the point $a \in X$ to a small neighborhood of the

---

[1] This is usually called a topological branch point (*translator's note*).

[2] This is usually called a topological branched map (*translator's note*).

origin on the complex line. Let U^* be the preimage of a small punctured disk D_r^* with center at 0 under the map u. Suppose that among all connected components of the preimage $\pi^{-1}(U^*)$, there exists a component V^* such that the restriction of the map π to V^* is a k-fold covering. In this case, one can apply the *real operation of filling a hole*. The operation does the following: Cut a neighborhood V^* out of the manifold M. The covering $\pi : V^* \to U^*$ is replaced by the map $\pi : V \to U$ by the operation of filling a hole described above. The manifold V^* lies in V and differs from V at one point. The real operation of filling a hole attaches the neighborhood V to the manifold $M \setminus V^*$ together with the map $\pi : V \to X$.

The *real operation of filling holes* consists of real operations of filling a hole applied to all holes simultaneously. It is well defined: if V^* is a connected component of the preimage $\pi^{-1}(U^*)$, where U^* is a punctured neighborhood of the point $o \in X$, and the map $\pi : V^* \to U^*$ is a finite covering, then the operation of filling all holes adds to the closure of the domain V^* exactly one point lying over the point o. The topology near this new point is defined in the same way as under the operation of filling one hole.

The *operation of filling holes* is the complexification of the real operation of filling holes. The operation of filling holes can be applied to a one-dimensional complex analytic manifold M endowed with an analytic map $f : M \to X$. Namely, the triple $f : M \to X$ should be regarded as an analytic-type map from a real manifold M to X. Then the real operation of filling holes should be applied to this triple. The result is a real manifold $\tilde{M}$ together with an analytic-type map $\pi : \tilde{M} \to X$. The manifold $\tilde{M}$ has a unique structure of a complex manifold such that the analytic-type map π is analytic. This complex manifold $\tilde{M}$ together with the analytic map π is the result of the operation of filling holes applied to the initial triple $f : M \to X$. In what follows, we will need only the operation of filling holes and not its real version.

Let X and M be one-dimensional complex manifolds, O a discrete subset of X, and $\pi : M \to U$, where $U = X \setminus O$, an analytic map that is a finite covering. Let X be connected (the covering space M may be disconnected).

Near every point $o \in O$, one can take a small punctured neighborhood U^* that does not contain other points of the set O. Over the punctured neighborhood U^*, there is a covering $f : V^* \to U^*$, where $V^* = f^{-1}(U^*)$. The manifold V^* splits into connected components V_i^*. Let us apply the operation of filling holes. Over the point $o \in O$, we attach a finite number of points. The number of points is equal to the number of connected components of V^*.

Lemma 2.2.6 *If the operation of filling holes is applied to a k-fold covering $\pi :$ $M \to U$, then the result is a complex manifold $\tilde{M}$ endowed with a proper analytic map $\tilde{\pi} : \tilde{M} \to X$ of degree k.*

Proof We should check the properness of the map $\tilde{\pi}$. First of all, this map is analytic, and hence the image of every open subset under this map is open. Next, the number of preimages of every point $x_0 \in X$ under the map $\tilde{\pi}$, counted with multiplicity, is equal to k. Hence, the map $\tilde{\pi}$ is proper.   □

2.2.3 Finite Ramified Coverings with a Fixed Ramification Set

In this subsection, we classify finite ramified coverings with a fixed ramification set.

Let X be a connected complex manifold with a distinguished discrete subset O. A triple consisting of complex manifolds M and X and a proper analytic map $\pi : M \to X$ whose critical values are all contained in the set O is called a *ramified covering over X with ramification over O*. We consider ramified coverings up to left equivalence. In other words, two triples $\pi_1 : M_1 \to X_1$ and $\pi_2 : M_2 \to X_2$ are considered the same if there exists a homeomorphism $h : M_1 \to M_2$ compatible with the projections π_1 and π_2, i.e., $\pi_1 = h \circ \pi_2$. *The homeomorphism h that establishes the equivalence of ramified coverings is automatically an analytic map from the manifold M_1 to the manifold M_2.* This is proved in the same way as Proposition 2.2.5.

The following operation will be called the *ramification puncture*. To every connected covering $\pi : M \to X$ ramified over O, the operation assigns the unramified covering $\pi : M \setminus \tilde{O} \to X \setminus O$ over $X \setminus O$, where $\tilde{O}$ is the full preimage of the set O under the map π. The following lemma is a direct consequence of the relevant definitions.

Lemma 2.2.7 *The operation of ramification puncture and the operation of filling holes are inverse to each other. They establish an isomorphism between the category of ramified coverings over X with ramifications over the set O and the category of finite coverings over $X \setminus O$.*

All definitions and statements about coverings can be extended to ramified coverings. This is done automatically: it is enough to apply arguments used in the proof of Proposition 2.2.5. Thus we formulate definitions and propositions about only ramified coverings.

Let us start with definitions concerning ramified coverings. A homeomorphism $h : M \to M$ is called a *deck transformation of a ramified covering $\pi : M \to X$ with ramification over O* if the equality $\pi = \pi \circ h$ is satisfied. (The deck transformation h is automatically analytic.)

For a connected manifold M, a ramified covering $\pi : M \to X$ with ramification over O is called *normal* if its group of deck transformations acts transitively on every fiber of the map π. The group of deck transformations is automatically a group of analytic transformations of M.

A ramified covering $f_2 : M_2 \to X$ with ramification over O is said to be *subordinate* to a normal ramified covering $f_1 : M_1 \to X$ with ramification over O if there exists a ramified covering $h : M_1 \to M_2$ with ramification over $f_2^{-1}(O)$ such that $f_1 = f_2 \circ h$. (The map h is automatically analytic.)

We now proceed to definitions concerning coverings with marked points. A triple $\pi : (M, b) \to (X, a)$, where $\pi : M \to X$ is a ramified covering with ramification over O, and $a \in X$, $b \in M$ are marked points such that $a \notin O$ and $\pi(b) = a$, is called a *ramified covering over X with marked points with ramification over O*.

A ramified covering $f_2 : (M_2, b_2) \to (X, a)$ with ramification over O is said to be *subordinate* to a ramified covering $f_1 : (M_1, b_1) \to (X, a)$ with ramification over O if there exists a ramified covering $h : (M_1, b_1) \to (M_2, b_2)$ with ramification over $f_2^{-1}(O)$ such that $f = f_2 \circ h$. (The map h is automatically analytic.) In particular, such coverings are called *equivalent* if the map h is a homeomorphism. (The homeomorphism h is automatically a bianalytic bijection between M_1 and M_2.)

The operation of ramification puncture assigns to a ramified covering $f : (Y, b) \to (X, a)$ with marked points and to a covering $\pi : M \to X$ with ramification over O the covering with marked points $f : (Y \setminus \pi^{-1}(O), b) \to (X \setminus O, a)$ and the covering $\pi : M \setminus \pi^{-1}(O) \to X \setminus O$. With these coverings over $X \setminus O$, one associates the subgroup of finite index in the group $\pi_1(X \setminus O, a)$ and, respectively, the class of conjugate subgroups of finite index in this group. We say that this subgroup corresponds to the ramified covering $f : (Y, b) \to (X, a)$ with marked points and that this class of conjugate subgroups corresponds to the ramified covering $\pi : M \to X$.

Consider all possible ramified coverings with marked points with a connected covering space over a manifold X with a marked point a that have ramification over a set O, $a \notin O$. Transferring the statements proved for coverings with marked points to ramified coverings, we obtain the following:

1. Such coverings are classified by subgroups of finite index in the group

$$\pi_1(X \setminus O, a).$$

2. Such a covering corresponding to the group G_2 is subordinate to the covering corresponding to the group G_1 if and only if the inclusion $G_2 \supseteq G_1$ holds.
3. Such a covering is normal if and only if the corresponding subgroup of the fundamental group $\pi_1(X \setminus O, a)$ is a normal subgroup H. The group of deck transformations of the normal ramified covering is isomorphic to

$$\pi_1(X \setminus O, a)/H.$$

Consider all possible ramified coverings over a manifold X with a connected covering space that have ramification over a set O, $a \notin O$. Transferring the statements proved for coverings to ramified coverings we obtain the following:

4. Such coverings are classified by conjugacy classes of subgroups of finite index in the group $\pi_1(X \setminus O, a)$.

One can literally translate the *description of ramified coverings subordinate to a given normal covering with the deck transformation group N* to ramified coverings. To a ramified covering with a marked point, assign the subgroup of the deck transformation group N that is equal to the image under the quotient projection $\pi_1(X, a) \to N$ of the subgroup of the fundamental group corresponding to the ramified covering. For this correspondence, we have the following theorem.

Theorem 2.2.8 *The correspondence between ramified coverings with marked points subordinate to a given normal covering and subgroups of the deck transformation group of this normal covering is bijective.*

Subordinate ramified coverings with marked points are equivalent as coverings if and only if the corresponding subgroups are conjugate in the deck transformation group.

A subordinate ramified covering is normal if and only if it corresponds to a normal subgroup M of the deck transformation group N. The deck transformation group of the subordinate normal covering is isomorphic to the quotient group N/M.

The notion of subcovering extends to ramified coverings. Let $f : M \rightarrow X$ be a normal ramified covering (as usual, we assume that the complex manifold M is connected). An *intermediate ramified covering* between M and X is a complex manifold Y together with a surjective analytic map $h_Y : M \rightarrow Y$ and a projection $f_Y : Y \rightarrow X$ satisfying the condition $f = f_Y \circ h_Y$.

We say that two intermediate ramified coverings

$$M \xrightarrow{\ h_1\ } Y_1 \xrightarrow{\ f_1\ } X \quad \text{and} \quad M \xrightarrow{\ h_2\ } Y_2 \xrightarrow{\ f_2\ } X$$

are equivalent as *ramified subcoverings of the covering* $f : M \rightarrow X$ if there exists a homeomorphism $h : Y_1 \rightarrow Y_2$ that makes the following diagram commutative:

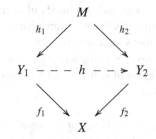

That is, the homeomorphism should be such that $h_2 = h \circ h_1$ and $f_1 = f_2 \circ h$. We say that two ramified subcoverings are equivalent as *ramified coverings over X* if there exists an analytic map $h : Y_1 \rightarrow Y_2$ such that $f_1 = h \circ f_2$ (the map h is not required to make the upper part of the diagram commutative).

The classification of intermediate ramified coverings regarded as ramified subcoverings is equivalent to the classification of subordinate coverings with marked points. Indeed, if we mark a point b in the manifold M that lies over the point a, then we obtain a canonically defined marked point $h_Y(a)$ in the space Y.

Let us reformulate Proposition 2.1.11.

Proposition 2.2.9 *Intermediate ramified coverings for a normal covering with the deck transformation group N:*

1. *regarded as ramified subcoverings are classified by subgroups of the group N;*
2. *regarded as ramified coverings over X are classified by the conjugacy classes of subgroups in the group N.*

A subordinate ramified covering is normal if and only if it corresponds to a normal subgroup H of the deck transformation group N. The deck transformation group of the subordinate ramified normal covering is isomorphic to the quotient group N/H.

Let us give one more description of ramified coverings subordinate to a given normal ramified covering. Let $\pi : M \to X$ be a normal finite ramified covering with the deck transformation group N.

The deck transformation group N is a group of analytic transformations of the manifold M commuting with the projection π. It induces a transitive transformation group of the fiber of π. Transformations in the group N can have isolated fixed points over the set of critical values of the map π.

Lemma 2.2.10 *The set M_N of orbits under the action of the deck transformation group N on a ramified normal covering M is in one-to-one correspondence with the manifold X.*

Proof By definition, deck transformations act transitively on the fiber of the map $\pi : M \to X$ over every point $x_0 \notin O$. Let $o \in O$ be a point in the ramification set. Let U^* be a small punctured coordinate disk around the point o not containing points of the set O. The preimage $\pi^{-1}(U^*)$ of the domain U^* splits into connected components V_i^* that are punctured neighborhoods of preimages b_i of the point o. The deck transformation group gives rise to a transitive permutation of the domains V_i^*. Indeed, each of these domains intersects the fiber $\pi^{-1}(c)$, where c is any point in the domain U^*, and the group N acts transitively on the fiber $\pi^{-1}(c)$. The transitivity of the action of N on the set of components V_i^* implies the transitivity of the action of N on the fiber $\pi^{-1}(o)$. $\square$

Theorem 2.2.11 *The set of orbits M/G of the analytic manifold M under the action of a finite group G of analytic transformations has a structure of an analytic manifold.*

Proof

1. The stabilizer G_{x_0} of every point $x_0 \in M$ under the action of the group G is cyclic. Indeed, consider the following homomorphism of the group G_{x_0} to the group of linear transformations of a one-dimensional vector space. The homomorphism assigns to the transformation its differential at the point x_0. This map cannot have a nontrivial kernel: if the first few terms of the Taylor series of the transformation f have the form $f(x_0 + h) = x_0 + h + ch^k + \cdots$, then the Taylor series of the lth iteration $f^{[l]}$ of f has the first few terms of the form $f^{[l]} = x_0 + h + lch^k + \cdots$. Hence, none of the iterations of the transformation f is the identity map, which contradicts the finiteness of the group G_{x_0}. A finite group of linear transformations of the space $\mathbb{C}^1$ is a cyclic group generated by multiplication by one of the primitive mth roots of unity ξ_m, where m is the order of the group G_{x_0}.

2. The stabilizer G_{x_0} of the point x_0 can be linearized, i.e., one can introduce a local coordinate u near x_0 such that the transformations in the group G_{x_0} written in this coordinate system are linear. Let f be a generator of the group G_{x_0}. Then the equality $f^{[m]} = \text{Id}$ holds, where Id is the identity transformation. The differential of the function f at the point x_0 is equal to multiplication by ξ_m, where ξ_m is one of the primitive roots of unity of order m. Consider any function φ whose differential is not equal to zero at the point x_0. With the map f, one associates the linear operator f^* on the space of functions. Let us write the Lagrange resolvent $R_{\xi_m}(\varphi)$ of the function φ for the action of the operator f^*: $R_{\xi_m}(\varphi) = \frac{1}{m} \sum \xi_m^{-k}(f^*)^k(\varphi)$. The function $u = R_{\xi_m}(\varphi)$ is the eigenvector of the transformation f^* with eigenvalue ξ_m. The differentials at the point x_0 of the functions u and φ coincide (this can be verified by a simple calculation). The map f becomes linear in the u-coordinate, since $f^*u = \xi_m u$.

3. We now introduce an analytic structure on the space of orbits. Consider any orbit. Suppose first that the stabilizers of the points in the orbit are trivial. Then a small neighborhood of the point in the orbit intersects each orbit at most once. A local coordinate near this point parameterizes neighboring orbits. If a point in the orbit has a nontrivial stabilizer, then we choose a local coordinate u near the point such that in this coordinate system, the stabilizer acts linearly, multiplying u by the powers of the root ξ_m. The neighboring orbits are parameterized by the function $t = u^m$. The theorem is proved. $\qquad\square$

With every subgroup G of the group N, we associate the analytic manifold M_G, which is the space of orbits under the action of the group G. Identify the manifold M_N with the manifold X. Under this identification, the quotient map $f_{e,N} : M \to M_N$ coincides with the original covering $f : M \to X$.

Let G_1, G_2 be two subgroups in N such that $G_1 \subseteq G_2$. Define the map $f_{G_1,G_2} : M_{G_1} \to M_{G_2}$ by assigning to each orbit of the group G_1 the orbit of the group G_2 that contains it. It is easy to see that the following statements hold:

1. The map f_{G_1,G_2} is a ramified covering.
2. If $G_1 \subseteq G_2 \subseteq G_3$, then $f_{G_1,G_2} = f_{G_2,G_3} f_{G_1,G_2}$.
3. Under the identification of M_N with X, the map $f_{G,N} : M_G \to M_N$ corresponds to a ramified covering subordinate to the original covering $f_{e,N} : M \to M_N$ (since $f_{e,N} = f_{G,N} \circ f_{e,G}$).
4. If G is a normal subgroup of N, then the ramified covering $f_{G,N} : M_G \to M_N$ is normal, and its deck transformation group is equal to N/G.

With an intermediate ramified covering $f_{G,N} : M_G \to M_N$ one can associate either the triple of spaces $M \xrightarrow{f_{e,G}} M_G \xrightarrow{f_{G,N}} M_N$ with the maps $f_{e,G}$ and $f_{G,N}$, or the pair of spaces $M_G \xrightarrow{f_{G,N}} M_N$ with the map $f_{G,N}$. This two possibilities correspond to two viewpoints with respect to an intermediate covering regarded either as a ramified subcovering or as a ramified covering over M_N.

2.2.4 Riemann Surface of an Algebraic Equation over the Field of Meromorphic Functions

Our goal is a geometric description of algebraic extensions of the field $K(X)$ of meromorphic functions on a connected one-dimensional complex manifold X. In this subsection, we construct the Riemann surface of an algebraic equation over the field $K(X)$.

Let $T = y^n + a_1 y^{n-1} + \cdots + a_n$ be a polynomial in the variable y over the field $K(X)$ of meromorphic functions on X. We will assume that in the factorization of T, every irreducible factor occurs with multiplicity one. In this case, the discriminant D of the polynomial T is a nonzero element of the field $K(X)$. Denote by O the discrete subset of X containing all poles of the coefficients a_i and all zeros of the discriminant D. For every point $x_0 \in X \setminus O$, the polynomial $T_{x_0} = y^n + a_1(x_0)y^{n-1} + \cdots + a_n(x_0)$ has exactly n distinct roots. The *Riemann surface* of the equation $T = 0$ is an n-fold ramified covering $\pi : M \to X$ together with a meromorphic function $y : M \to \mathbb{C}P^1$ such that for every point $x_0 \in X \setminus O$, the set of roots of the polynomial T_{x_0} coincides with the set of values of the function y on the preimage $\pi^{-1}(x_0)$ of the point x_0 under the projection π. Let us show that there exists a unique Riemann surface of the equation (up to an analytic homeomorphism compatible with the projection to X and with the function y).

For the Cartesian product $(X \setminus O) \times \mathbb{C}^1$, one defines the projection π onto the first factor and the function y as the projection onto the second factor. Consider the hypersurface M_O in the Cartesian product given by the equation $T(\pi(a), y(a)) = 0$. The partial derivative of T with respect to y (with respect to the second argument) at every point of the hypersurface M_O is nonzero, since the polynomial $T_{\pi(a)}$ has no multiple roots. By the implicit function theorem, the hypersurface M_O is nonsingular, and its projection onto $X \setminus O$ is a local homeomorphism. The projection $\pi : M_O \to X \setminus O$ and the function $y : M_O \to \mathbb{C}^1$ are defined on the manifold M_O. Applying the operation of filling holes to the covering $\pi : M_O \to X \setminus O$, we obtain an n-fold ramified covering $\pi : M \to X$.

Theorem 2.2.12 *The function $y : M_O \to \mathbb{C}^1$ can be extended to a meromorphic function $y : M \to \mathbb{C}P^1$. The ramified covering $\pi : M \to X$ endowed with the meromorphic function $y : M \to \mathbb{C}P^1$ is the Riemann surface of the equation $T = 0$. There are no other Riemann surfaces of the equation $T = 0$.*

Proof We need the following lemma.

Lemma 2.2.13 (From high-school mathematics) *Every root y_0 of the equation $y^n + a_1 y^{n-1} + \cdots + a_n = 0$ satisfies the inequality $|y_0| \le \max(1, \sum |a_i|)$.*

Proof If $|y_0| > 1$ and $y_0 = -a_1 - \cdots - a_n y_0^{1-n}$, then $|y_0| \le \max(1, \sum |a_i|)$. $\square$

Let us now prove the theorem. The functions $\pi^* a_i$ are meromorphic on M. In the punctured neighborhood of every point, the function y satisfies the inequality

$|y| \leq \max(1, \sum |\pi^* a_i|)$ and therefore has a pole or a removable singularity at every added point.

By construction, the triple $\pi : M_O \to X \setminus O$ is an n-fold covering, and for every $x_0 \in X \setminus O$, the set of roots of the polynomial T_{x_0} coincides with the image of the set $\pi^{-1}(x_0)$ under the map $y : M_O \to \mathbb{C}P^1$. Therefore, the ramified covering $\pi : M \to X$ endowed with the meromorphic function $y : M \to \mathbb{C}P^1$ is the Riemann surface of the equation $T = 0$.

Let a ramified covering $\pi_1 : M_1 \to X_1$ endowed with the function $y : M_1 \to \mathbb{C}P^1$ be another Riemann surface of this equation. Let O_1 denote the set $\pi_1^{-1} O$. There exists a natural bijective map $h_1 : M_O \to M_1 \setminus O_1$ such that $\pi_1 \circ h_1 = \pi$ and $y_1 \circ h_1 = y$. Indeed, by definition of the Riemann surface, the sets of numbers $\{y \circ \pi^{-1}(x)\}$ and $\{y_1 \circ \pi_1^{-1}(x)\}$ coincide with the set of roots of the polynomial $T_{\pi(x)}$. It is easy to see that the map h_1 is continuous and that it can be extended by continuity to an analytic homeomorphism $h : M \to M$ such that $\pi_1 \circ h = \pi$ and $y_1 \circ h = y$. The theorem is proved. $\qquad\square$

Remark 2.2.14 Sometimes the manifold M in the definition of the Riemann surface of an equation is itself called the Riemann surface of an equation. The same manifold is called the *Riemann surface of the function y satisfying the equation*. We will use this slightly ambiguous terminology whenever this does not lead to confusion.

The set $\tilde{O}$ of critical values of the ramified covering $\pi : M \to X$ associated with the Riemann surface of the equation $T = 0$ can be strictly contained in the set O used in the construction (the inclusion $\tilde{O} \subset O$ always holds). The set $\tilde{O}$ is called the *ramification set of the equation $T = 0$*. Over a point $a \in X \setminus \tilde{O}$, the equation $T_a = 0$ might have multiple roots. However, in the field of germs of meromorphic functions at the point $a \in X \setminus \tilde{O}$, the equation $T = 0$ has only nonmultiple roots, and their number is equal to the degree of the equation $T = 0$. *Each of the meromorphic germs at the point a satisfying the equation $T = 0$ corresponds to a point over a in the Riemann surface of the equation.*

Chapter 3
Ramified Coverings and Galois Theory

This chapter is based on Galois theory and the Riemann existence theorem (which we accept without proof) and is devoted to the relationship between finite ramified coverings over a manifold X and algebraic extensions of the field $K(X)$. For a finite ramified covering M, we show that the field $K(M)$ of meromorphic functions on M is an algebraic extension of the field $K(X)$ of meromorphic functions on X, and that every algebraic extension of the field K can be obtained in this way.

The following construction plays a key role. Fix a discrete subset O of a manifold X and a point $a \in X \setminus O$. Consider the field $P_a(O)$ consisting of the meromorphic germs at the point $a \in X$ that can be meromorphically continued to multivalued functions on $X \setminus O$ with finitely many branches and with algebraic singularities at the points of the set O. The operation of meromorphic continuation of a germ along a closed curve gives the action of the fundamental group $\pi_1(X \setminus O)$ on the field $P_a(O)$. The results of Galois theory are applied to the action of this group of automorphisms of the field $P_a(O)$. We describe the correspondence between subfields of the field $P_a(O)$ that are algebraic extensions of the field $K(X)$ and the subgroups of finite index of the fundamental group $\pi_1(X \setminus O)$. We prove that this correspondence is bijective. Apart from Galois theory, the proof uses the Riemann existence theorem.

Normal ramified coverings over a connected complex manifold X are connected with Galois extensions of the field $K(X)$. The main theorem of Galois theory for such extensions has a transparent geometric interpretation.

The local variant of the connection between ramified coverings and algebraic extensions allows one to describe algebraic extensions of the field of convergent Laurent series. Extensions of this field are analogous to algebraic extensions of the finite field $\mathbb{Z}/p\mathbb{Z}$ (under this analogy, the homotopy class of closed real curves passing around the point 0 corresponds to the Frobenius automorphism).

At the end of the chapter, compact one-dimensional complex manifolds are considered. On the one hand, arguments of Galois theory show that the field of meromorphic functions on a compact manifold is a finitely generated extension of the field of complex numbers of transcendence degree one (the proof uses the Riemann existence theorem). On the other hand, ramified coverings allow one to describe

A. Khovanskii, *Galois Theory, Coverings, and Riemann Surfaces*, DOI 10.1007/978-3-642-38841-5_3, © Springer-Verlag Berlin Heidelberg 2013

explicitly enough all algebraic extensions of the field of rational functions in one variable. Consider the extension obtained by adjoining all roots of a given algebraic equation. The Galois group of such an extension has a geometric meaning: it coincides with the monodromy group of the Riemann surface of the algebraic function defined by this equation. Hence, Galois theory yields a topological obstruction to the representability of algebraic functions in terms of radicals.

3.1 Finite Ramified Coverings and Algebraic Extensions of Fields of Meromorphic Functions

Let $\pi : M \to X$ be a finite ramified covering. Galois theory and the Riemann existence theorem allow one to describe the connection between the field $K(M)$ of meromorphic functions on M and the field $K(X)$ of meromorphic functions on X. The field $K(M)$ is an algebraic extension of the field $K(X)$, and every algebraic extension of the field $K(X)$ can be obtained in this way. This section is devoted to the connection between finite ramified coverings over the manifold X and algebraic extensions of the field $K(X)$.

In Sect. 3.1.1, we define the field $P_a(O)$ consisting of the meromorphic germs at the point $a \in X$ that can be meromorphically continued to multivalued functions on $X \setminus O$ with finitely many branches and with algebraic singularities at the points of the set O.

In Sect. 3.1.2, the action of the fundamental group $\pi_1(X \setminus O)$ on the field $P_a(O)$ is considered and the results of Galois theory are applied to the action of this group of automorphisms. We describe the correspondence between subfields of the field $P_a(O)$ that are algebraic extensions of the field $K(X)$ and the subgroups of finite index in the fundamental group $\pi_1(X \setminus O)$. We prove that this correspondence is bijective (apart from Galois theory, the proof uses the Riemann existence theorem). Consider the Riemann surface of an equation whose ramification set lies over the set O. We show that this Riemann surface is connected if and only if the equation is irreducible. The field of meromorphic functions on the Riemann surface of an irreducible equation coincides with the algebraic extension of the field $K(X)$ obtained by adjoining a root of the equation.

In Sect. 3.1.3, we show that the field of meromorphic functions on every connected ramified finite covering of X is an algebraic extension of the field $K(X)$ and different extensions correspond to different coverings.

3.1.1 The Field $P_a(O)$ of Germs at the Point $a \in X$ of Algebraic Functions with Ramification over O

Let X be a connected complex manifold, O a discrete subset of X, and a a marked point in X not belonging to the set O.

Let $P_a(O)$ denote the collection of germs of meromorphic functions at the point a with the following properties. A germ φ_a lies in $P_a(O)$ if

1. the germ φ_a can be extended meromorphically along any path that originates at the point a and lies in $X \setminus O$;
2. for the germ φ_a, there exists a subgroup $G_0 \subset \pi_1(X \setminus O, a)$ of finite index in the group $\pi_1(X \setminus O, a)$ such that under the continuation of the germ φ_a along a path in the subgroup G_0, one obtains the initial germ φ_a;
3. the multivalued analytic function on $X \setminus O$ obtained by analytic continuation of the germ φ_a has algebraic singularities at the points of the set O.

Let us discuss property 3 in more detail. Let $\gamma : [0, 1] \to X$ be any path that goes from the point a to a singular point $o \in O$, $\gamma(0) = a$, $\gamma(1) = o$, inside the domain $X \setminus O$, that is, $\gamma(t) \in X \setminus O$ if $t < 1$. Property 3 means the following. For all values of the parameter t sufficiently close to 1 ($t_0 < t < 1$), consider the germs obtained by analytic continuation of φ_a along the path γ up to the point $\gamma(t)$. These germs are analytic, and they define a k-valued analytic function φ_γ in a small punctured neighborhood V_o^* of the point o. The restriction of the function φ_γ to a small punctured coordinate disk $D_{|u| < r}^*$ with center at the point o, where u is a local coordinate near the point o such that $u(o) = 0$, must have an algebraic singularity in the sense of the definition from Sect. 2.2.1. The last condition does not depend on the choice of a coordinate function u. It means that the function φ_γ can be expanded into a Puiseux series in u (or equivalently, the function grows no faster than a power of u when approaching the point o).

Lemma 3.1.1 *The set of germs $P_a(O)$ is a field. The fundamental group G of the domain $X \setminus O$ acts on the field $P_a(O)$ by analytic continuation. The invariant subfield of this action is the field of meromorphic functions on the manifold X.*

Proof Suppose that the germs $\varphi_{1,a}$ and $\varphi_{2,a}$ lie in the field $P_a(O)$ and do not change under continuations along the subgroups G_1 and G_2 of finite index in the group $G = \pi_1(X \setminus O, a)$. Then the germs $\varphi_{1,a} \pm \varphi_{2,a}$, $\varphi_{1,a}\varphi_{2,a}$, and $\varphi_{1,a}/\varphi_{2,a}$ (the germ $\varphi_{1,a}/\varphi_{2,a}$ is well defined, provided that the germ $\varphi_{2,a}$ is not identically equal to zero) can be extended meromorphically along any path that originates at the point a and lies in the domain $X \setminus O$. These germs do not change under continuation along the subgroup $G_1 \cap G_2$ of finite index in the group $G = \pi_1(X \setminus O, a)$.

Multivalued functions defined by these germs have algebraic singularities at the points of the set O, since the germs of functions representable by Puiseux series form a field. (Of course, one cannot apply arithmetic operations to multivalued functions. However, for a fixed path passing through the point 0, one can apply arithmetic operations to the fixed branches of the functions representable by Puiseux series. As a result, one obtains a branch of the function representable by a Puiseux series.)

Thus we have shown that $P_a(O)$ is a field. Meromorphic continuation preserves arithmetic operations. Therefore, the fundamental group G acts on $P_a(O)$ by automorphisms. The invariant subfield consists of the germs in the field $P_a(O)$ that are the germs of meromorphic functions in the domain $X \setminus O$. At the points of the set O, these single-valued functions have algebraic singularities and therefore are meromorphic functions on the manifold X. The lemma is proved. $\qquad\square$

3.1.2 Galois Theory for the Action of the Fundamental Group on the Field $P_a(O)$

In this subsection, we will apply Galois theory to the action of the fundamental group $G = \pi_1(X \setminus O, a)$ on the field $P_a(O)$.

Theorem 3.1.2 *The following properties hold:*

1. *Every element φ_a of the field $P_a(O)$ is algebraic over the field $K(X)$.*
2. *The set of germs at the point a satisfying the same irreducible equation as the germ φ_a coincides with the orbit of the germ φ_a under the action of the group G.*
3. *The germ φ_a lies in the field obtained by adjoining an element f_a of the field $P_a(O)$ to the field $K(X)$ if and only if the stabilizer of the germ φ_a under the action of the group G contains the stabilizer of the germ f_a.*

Proof The proof of parts 1 and 2 follow from Theorem 1.3.3, and the proof of part 3 follows from Theorem 1.3.9. $\qquad\square$

Part 1 of Theorem 3.1.2 can be reformulated as follows.

Proposition 3.1.3 *A meromorphic germ at the point a lies in the field $P_a(O)$ if and only if it satisfies an irreducible equation $T = 0$ whose set of ramification points is contained in the set O.*

Part 2 of Theorem 3.1.2 is equivalent to the following statement.

Proposition 3.1.4 *Consider an equation $T = 0$ whose set of ramification points is contained in the set O. The equation T is irreducible if and only if the Riemann surface of this equation is connected.*

Proof Let $f : M \to X$ be a Riemann surface of an equation whose set of ramification points is contained in the set O. By part 2 of Theorem 3.1.2, the equation is irreducible if and only if the manifold $M \setminus f^{-1}(O)$ is connected. Indeed, the connectedness of the covering space is equivalent to the fact that the fiber $F = f^{-1}(a)$ lies in a single connected component of the covering space. This, in turn, implies the transitivity of the action of the monodromy group on the fiber F. It remains to note that the manifold M is connected if and only if the manifold $M \setminus f^{-1}(O)$ obtained by removing a discrete subset from M is also connected. $\qquad\square$

Proposition 3.1.5 *A subfield of the field $P_a(O)$ is a normal extension of the field $K(X)$ if and only if it is obtained by adjoining all germs at the point a of a multivalued function on X satisfying an irreducible algebraic equation $T = 0$ over X whose ramification lies over O. The Galois group of this normal extension is isomorphic to the monodromy group of the Riemann surface of the equation $T = 0$.*

Proof A normal extension is always obtained by adjoining all roots of an irreducible equation. In the setting of the proposition, the ramification set of this equation must be contained in O. Both the Galois group of the normal covering and the monodromy group of the equation $T = 0$ are isomorphic to the image of the fundamental group $\pi_1(X \setminus O, a)$ under its action on the orbit in the field $P_a(O)$ consisting of those germs at the point a that satisfy the equation $T = 0$. $\square$

Consider the Riemann surface of the equation $T = 0$ whose root is a germ $\varphi_a \in P_a(O)$. The points of this Riemann surface lying over the point a correspond to the roots of the equation $T = 0$ in the field $P_a(O)$. The germ φ_a is one of these roots. In this way, we assigned to each germ φ_a of the field $P_a(O)$ firstly, the ramified covering $\pi_{\varphi_a} : M_{\varphi_a} \to X$ whose set of critical values is contained in O, and secondly, the marked point $\varphi_a \in M_{\varphi_a}$ lying over the point a (the symbol φ_a denotes the point of the Riemann surface corresponding to the germ φ_a). Part 3 of Theorem 3.1.2 can be reformulated as follows.

Proposition 3.1.6 *A germ φ_a lies in the field obtained by adjoining an element f_a of the field $P_a(O)$ to the field $K(X)$ if and only if the ramified covering $\pi_{\varphi_a} : (M_{\varphi_a}, \varphi_a) \to (X, a)$ is subordinate to the ramified covering $\pi_{f_a} : (M_{f_a}, f_a) \to (X, a)$.*

Indeed, according to the classification of ramified coverings with marked points, the covering corresponding to the germ φ_a is subordinate to the covering corresponding to the germ f_a if and only if the stabilizer of the germ φ_a under the action of the fundamental group $\pi_1(X \setminus O)$ contains the stabilizer of the germ f_a.

Corollary 3.1.7 *The fields obtained by adjoining elements φ_a and f_a of the field $P_a(O)$ to the field $K(X)$ coincide if and only if the ramified coverings with marked points $\pi_{\varphi_a} : (M_{\varphi_a}, \varphi_a) \to (X, a)$ and $\pi_{f_a} : (M_{f_a}, f_a) \to (X, a)$ are equivalent.*

Is this true that for every subgroup G of finite index in the fundamental group $\pi_1(X \setminus O, a)$, there exists a germ $f_a \in P_a(O)$ whose stabilizer is equal to G?

The answer to this question is positive. Galois theory alone does not suffice to prove this fact: in order to apply algebraic arguments, we need to have plenty of meromorphic functions on the manifold.[1] It will be sufficient for us to use the fact

---

[1]Galois theory allows one to obtain the following result. Suppose that the answer for a subgroup G is positive, and let $f_a \in P_a(O)$ be a germ whose stabilizer is equal to G. Let H denote the largest normal subgroup lying in G. Then for every subgroup containing the group H, the answer is also

formulated below, which we will call the Riemann existence theorem and apply
without proof. (The proof uses functional analysis and is not algebraic. Note that
there exist two-dimensional compact complex analytic manifolds such that the only
meromorphic functions on these manifolds are constants.)

Theorem 3.1.8 (Riemann existence theorem) *For every finite subset of a one-
dimensional analytic manifold, there exists a meromorphic function on this manifold, analytic in a neighborhood of the subset and taking different values at different
points of the subset.*

Theorem 3.1.9 *For every subgroup G of finite index in the fundamental group
$\pi_1(X \setminus O, a)$, there exists a germ $f_a \in P_a(O)$ whose stabilizer is equal to G.*

Proof Let $\pi : (M, b) \to (X, a)$ be a finite ramified covering over X whose critical
points lie over O. Assume that the covering corresponds to a subgroup $G \subset \pi_1(X \subset
O)$. Let $F = \pi^{-1}(a)$ denote the fiber of the covering over the point a. By the Rie-
mann existence theorem (Theorem 3.1.8), there exists a meromorphic function on
the manifold M that takes different values at different points of the set F. Let $\pi_{b,a}^{-1}$
be a germ of the inverse map to the projection π that takes the point a to a point b.
The germ of the function $f \circ \pi_{b,a}^{-1}$ lies in the field $P_a(O)$ by construction, and its
stabilizer under the action of the fundamental group $\pi_1(X \subset O)$ is equal to G. $\square$

Hence, we have shown that the classification of the algebraic extensions of the
meromorphic function field $K(X)$ that are contained in the field $P_a(O)$ is equivalent to the classification the of ramified finite coverings $\pi : (M, b) \to (X, a)$ whose
critical values lie in the set O. Both types of objects are classified by subgroups
of finite index in the fundamental group $\pi_1(X \setminus O, a)$. In particular, we have the
following theorem.

Theorem 3.1.10 *There is a bijective correspondence between subgroups of finite
index in the fundamental group and algebraic extensions of the field $K(X)$ that are
contained in the field $P_a(O)$. If a subgroup G_1 lies in the subgroup G_2, then the field
corresponding to the subgroup G_2 lies in the field corresponding to the subgroup
G_1. A subfield of $P_a(O)$ is a Galois extension of the field $K_a(X)$ if it corresponds to
a normal subgroup H of the fundamental group. The Galois group of this extension
is isomorphic to the quotient group $\pi_1(X \setminus O, a)/H$.*

---

positive. For the proof, it suffices to apply the fundamental theorem of Galois theory to the minimal
Galois extension of the field $K(X)$ containing the germ f_1.

3.1.3   Field of Functions on a Ramified Covering

Here we show that irreducible algebraic equations over the field $K(X)$ give rise to isomorphic extensions of this field if and only if the Riemann surfaces of these equations provide equivalent ramified coverings over the manifold X.

Proposition 3.1.4 implies the following corollary.

Corollary 3.1.11   *An algebraic extension of a field $K(X)$ is irreducible if and only if its Riemann surface is connected.*

Let $\pi : (M, b) \to (X, a)$ be a finite ramified covering with marked points such that the manifold M is connected and the point a does not belong to the set of critical values of the map π. We can apply the results about the field $P_a(O)$ and its subfields to describe the field of meromorphic functions on M. The following construction is useful.

Let $\pi_{b,a}^{-1}$ denote a germ of the inverse map to the projection π that takes the point a to a point b. Let $K_b(M)$ be the field of germs at the point b of meromorphic functions on the manifold M. This field is isomorphic to the field $K(M)$. The map $(\pi_{b,a}^{-1})^*$ embeds the field $K_b(M)$ in the field $P_a(O)$. Taking different preimages b of the point a, we obtain different embeddings of the field $K_b(M)$ in the field $P_a(O)$.

Suppose that an equation $T = 0$ is irreducible over the field $K(X)$. Then its Riemann surface is connected, and the meromorphic functions on this surface form the field $K(M)$. The field $K(M)$ contains the subfield $\pi^*(K(X))$ isomorphic to the field of meromorphic functions on the manifold M. Let $y : M \to \mathbb{C}P^1$ be a meromorphic function that appears in the definition of the Riemann surface. We have the following proposition.

Proposition 3.1.12   *The field $K(M)$ of meromorphic functions on the surface M is generated by the function y over the subfield $\pi^*(K(X))$. The function y satisfies the irreducible algebraic equation $T = 0$ over the subfield $\pi^*(K(X))$.*

Proof Let $b \in M$ be a point of the manifold M that is projected to the point a, $\pi(b) = a$, and let $\pi_{b,a}^{-1}$ be a germ of the inverse map to the projection π that takes the point a to a point b. Let $K_b(M)$ denote the field of germs at the point b of meromorphic functions on the manifold M. This field is isomorphic to the field $K(M)$. The map $(\pi_{b,a}^{-1})^*$ embeds the field $K_b(M)$ into the field $P_a(O)$.

For every meromorphic function $g : M \to \mathbb{C}P^1$, the germ $g_b \circ \pi_{b,a}^{-1}$ lies in the field $P_a(O)$. The stabilizer of this germ under the action of the group $\pi_1(X \setminus O, a)$ contains the stabilizer of the point b under the action of the monodromy group. For the germ $y_b \circ \pi_{b,a}^{-1}$, the stabilizer is equal to the stabilizer of the point b under the action of the monodromy group, since the function y by definition takes distinct values at the points of the fiber $\pi^{-1}(a)$. The proposition now follows from part 2 of Theorem 3.1.2. $\square$

Theorem 3.1.13 *Irreducible equations $T_1 = 0$ and $T_2 = 0$ over the field $K(X)$ give isomorphic extensions of this field if and only if the ramified coverings $\pi_1 : M_1 \to X$ and $\pi_2 : M_2 \to X$ that occur in the definition of the Riemann surfaces of these equations are equivalent.*

Proof Consider the points of the Riemann surfaces of the equations $T_1 = 0$ and $T_2 = 0$ that lie over a point x of the manifold X. For almost every x, these points are uniquely defined by the values of the roots y_1 and y_2 of the equations $T_1 = 0$ and $T_2 = 0$ over the point x. If the equations $T_1 = 0$ and $T_2 = 0$ define the same extensions of the field $K(X)$, then we have $y_1 = Q_1(y_2)$ and $y_2 = Q_2(y_1)$, where Q_1 and Q_2 are polynomials with coefficients from the field $K(X)$. These polynomials define almost everywhere an invertible map of one Riemann surface to the other that is compatible with projections of these surfaces to X. By continuity, it extends to an isomorphism of coverings.

If the Riemann surfaces of the equations give rise to equivalent coverings and a map $h : M_1 \to M_2$ establishes the equivalence, then h is compatible with the projections, and hence is analytic. The map $h^* : K(M_2) \to K(M_1)$ establishes the isomorphism of the fields $K(M_1)$ and $K(M_2)$ and takes the subfield $\pi_2^*(K(X))$ to the subfield $\pi_1^*(K(X))$, since $\pi_1 = \pi_2 \circ h$. $\qquad\Box$

3.2 Geometry of Galois Theory for Extensions of a Field of Meromorphic Functions

In this subsection, we put together the previous results. In Sect. 3.2.1 we discuss the relation between normal ramified coverings over a connected complex manifold X and Galois extensions of the field $K(X)$. In Sect. 3.2.2, this relation is used to describe extensions of the field of converging Laurent series.

In Sect. 3.2.3, we talk about complex one-dimensional manifolds. Galois theory helps to describe the field of meromorphic functions on a compact manifold, and geometry of ramified coverings allows us to describe explicitly enough all algebraic extensions of the field of rational functions in one variable.

The Galois group of an extension of the field of rational functions coincides with the monodromy group of the Riemann surface of an algebraic function defining this extension. Therefore, Galois theory gives a topological obstruction to the representability of algebraic functions in radicals.

3.2.1 Galois Extensions of the Field $K(X)$

By Theorem 3.1.13, algebraic extensions of the field of rational functions on a connected complex manifold X have a transparent geometric classification that coincides with the classification of connected finite ramified coverings over the manifold X. By this classification, Galois extensions of the field $K(X)$ correspond to normal ramified coverings over the manifold X. Let us describe all intermediate extensions for such Galois extensions.

Let X be a connected complex analytic manifold, $\pi : M \to X$ a normal ramified finite covering over X; also, let O be a finite subset of X containing all critical values of the map π, and $a \in X$ any point not in O. We have a field $K(X)$ of meromorphic functions on the manifold X and a Galois extension of this field, namely, the field $K(M)$ of meromorphic functions on the manifold M.

By Proposition 2.2.9, intermediate ramified coverings $M \xrightarrow{x_1} Y_1 \xrightarrow{f_1} X$ are in one-to-one correspondence with the subgroups of the deck transformation group N of the normal covering $\pi : M \to X$. With every ramified covering $M \xrightarrow{x_1} Y_1 \xrightarrow{f_1} X$ one can associate the subfield $x_1^*(K(Y_1))$ of the field $K(M)$ of meromorphic functions on the manifold M. As follows from the fundamental theorem of Galois theory, every intermediate field between $K(M)$ and $\pi^* K(X)$ is of this form, i.e., it is the field $x^*(K(Y))$ for an intermediate ramified covering $M \xrightarrow{x} Y \xrightarrow{f} X$. By this classification, intermediate Galois extensions of the field $K(X)$ correspond to intermediate normal coverings $M \xrightarrow{x} Y \xrightarrow{f} X$, and the Galois groups of intermediate Galois extensions are equal to the deck transformation groups of intermediate normal coverings.

Here is a slightly different description of the same Galois extension. The finite deck transformation group N acts on a normal ramified covering M. With each subgroup G of the group N, one can associate the subfield $K_N(M)$ of meromorphic functions on M invariant under the action of the group G.

Proposition 3.2.1 *The field $K(M)$ is a Galois extension of the field $K_N(M) = \pi^*(K(X))$. The Galois group of this Galois extension is equal to N. Under the Galois correspondence, a subgroup $G \subset N$ corresponds to the field $K_G(M)$.*

3.2.2 Algebraic Extensions of the Field of Germs of Meromorphic Functions

In this subsection, the relation between normal coverings and Galois extensions is used to describe extensions of the field of converging Laurent series.

Let L_0 be the field of germs of meromorphic functions at the point $0 \in \mathbb{C}^1$. This field can be identified with the field of convergent Laurent series $\sum_{m > m_0} c_m x^m$.

Theorem 3.2.2 *For every k, there exists a unique extension of degree k of the field L_0. It is generated by the element $z = x^{1/k}$. This extension is normal, and its Galois group is equal to $\mathbb{Z}/k\mathbb{Z}$.*

Proof Let $y^k + a_{k-1} y^{k-1} + \cdots + a_0 = 0$ be an irreducible equation over the field L_0. The irreducibility of the equation implies the existence of a small open disk D_r with center at the point 0 satisfying the following conditions:

1. All Laurent series a_i, $i = 1, \ldots, k$, converge in the punctured disk D_r^*.
2. The equation is irreducible over the field $K(D_r)$ of meromorphic functions in the disk D_r.
3. The discriminant of the equation does not vanish at any point of the punctured disk D_r^*.

Let $\pi : M \to D_r$ be the Riemann surface of the irreducible equation over the disk D_r. By the assumption, the point 0 is the only critical value of the map π. The fundamental group of the punctured disk D_r^* is isomorphic to the additive group of integers $\mathbb{Z}$. The group $k\mathbb{Z}$ is the only subgroup of index k in the group $\mathbb{Z}$. This subgroup is a normal subgroup, and the quotient group $\mathbb{Z}/k\mathbb{Z}$ is the cyclic group of order k. Hence, there exists a unique extension of degree k. It corresponds to the germ of a k-fold covering $f : (\mathbb{C}^1, 0) \to (\mathbb{C}^1, 0)$, where $f = z^k$. The extension is normal, and its Galois group is equal to $\mathbb{Z}/k\mathbb{Z}$. Next, the function $z : D_q \to \mathbb{C}^1$, where $q = r^{1/k}$, takes distinct values on all preimages of the point $a \in D_r^*$ under the map $x = z^k$. Hence, the function $z = x^{1/k}$ generates the field $K(D_q)$ over the field $K(D_r)$. The theorem is proved. $\square$

By the theorem, the function z and its powers $1, z = x^{1/k}, \ldots, z^{k-1} = x^{(k-1)/k}$ form a basis in the extension L of degree k of the field L_0 regarded as a vector space over the field L_0. Functions $y \in L$ can be regarded as multivalued functions of x. The expansion $y = f_0 + f_1 z + \cdots + f_{k-1} z^{k-1}$, $f_0, \ldots, f_{k-1} \in L_0$, of the element $y \in L$ in the given basis is equivalent to the expansion of the multivalued function $y(x)$ into the Puiseux series $y(x) = f_0(x) + f_1(x)x^{1/k} + \cdots + f_{k-1}(x)x^{(k-1)/k}$.

Note that the elements $1, z, \ldots, z^{k-1}$ are the eigenvectors of the following automorphism of the field L over the field L_0. The automorphism is defined by the analytic continuation along the loop around the point 0. It generates the Galois group. The eigenvalues of the given eigenvectors are equal to $1, \xi, \ldots, \xi^{k-1}$, where ξ is a primitive kth root of unity. The existence of such a basis of eigenvectors is proved in Galois theory (see Proposition 1.1.1).

Remark 3.2.3 The field L_0 is in many respects similar to the finite field $\mathbb{Z}/p\mathbb{Z}$. Continuation along the loop around the point 0 is similar to the Frobenius automorphism. Indeed, each of these fields has a unique extension of degree k for every positive integer k. All these extensions are normal, and their Galois groups are isomorphic to the cyclic group of k elements. The generator of the Galois group of the first field corresponds to a loop around the point 0, and the generator of the Galois group of the second field is the Frobenius automorphism. Every finite field has similar properties. For the field F_q consisting of $q = p^n$ elements, the role of a loop around the point 0 is played by the nth iterate of the Frobenius automorphism.

3.2.3 Algebraic Extensions of the Field of Rational Functions

Let us now consider the case of connected compact complex manifolds. Using Galois theory, we show that the field of meromorphic functions on such a manifold is a finite extension of transcendence degree one of the field of complex numbers. On the other hand, the geometry of the ramified coverings over the Riemann sphere provides a clear description of all finite algebraic extensions of the field of rational functions.

The Riemann sphere $\mathbb{C}P^1$ is the simplest of all compact complex manifolds. On the projective line $\mathbb{C}P^1$, we fix the point ∞ at infinity, $\mathbb{C}^1 \cap \{\infty\} = \mathbb{C}P^1$, and a holomorphic coordinate function $x : \mathbb{C}P^1 \to \mathbb{C}P^1$ that has a pole of order one at the point ∞. Every meromorphic function on $\mathbb{C}P^1$ is a rational function of x.

We say that a pair of meromorphic functions f, g on a manifold M *separates almost all points* of the manifold M if there exists a finite set $A \subset M$ such that the vector function (f, g) is defined on the set $M \setminus A$ and takes distinct values at all points of $M \setminus A$.

Theorem 3.2.4 *Let M be a connected compact one-dimensional complex manifold.*

1. *Then every pair of meromorphic functions f, g on M are related by a polynomial relation (i.e., there exists a polynomial Q in two variables such that the identity $Q(f, g) \equiv 0$ holds).*
2. *Let functions f, g separate almost all points of the manifold M. Then every meromorphic function φ on the manifold M is the composition of a rational function R in two variables with the functions f and g, that is, $\varphi = R(f, g)$.*

Proof

1. If the function f is identically equal to a constant C, then one can take the relation $f \equiv C$ as a polynomial relation. Otherwise, the map $f : M \to \mathbb{C}P^1$ is a ramified covering with a certain subset O of ramification points. It remains to use part 1 of Theorem 3.1.2.
2. If the function f is identically equal to a constant C, then the function g takes distinct values at the points of the set $M \setminus A$. Therefore, the ramified covering $g : M \to \mathbb{C}P^1$ is a bijective map of the manifold M to the Riemann sphere $\mathbb{C}P^1$. In this case, every meromorphic function φ on M is the composition of a rational function R in one variable with the function g, that is, $\varphi = R(g)$.

   If the function f is not constant, then it gives rise to the ramified covering $f : M \to \mathbb{C}P^1$ over the Riemann sphere $\mathbb{C}P^1$. Let O be the union of the set $f(A)$ and the set of critical values of the map f, let $a \in O$ be a point of the Riemann sphere not lying in O, and let F denote the fiber of the ramified covering $f : M \to \mathbb{C}P^1$ over the point a. By our assumption, the function f must separate the points of the set F. It remains to use part 3 of Theorem 3.1.2. $\square$

Let

$$y^n + a_{n-1}y^{n-1} + \cdots + a_0 \tag{1}$$

be an irreducible equation over the field of rational functions. The Riemann surface $\pi : M \to \mathbb{C}P^1$ of this equation is also called the *Riemann surface of the algebraic function* defined by this equation. The monodromy group of the ramified covering $\pi : M \to \mathbb{C}P^1$ is also called the *monodromy group of this algebraic function*. By Proposition 3.1.5, the Galois group of Eq. (1) coincides with the monodromy group.

Hence the Galois group of the irreducible Eq. (1) over the field of rational functions has a topological meaning: it is equal to the monodromy group of the Riemann surface of the algebraic function defined by Eq. (1). This fact was known to Frobenius, but it was probably discovered even earlier.

The results of Galois theory yield a topological obstruction to the solvability of Eq. (1) by radicals and k-radicals. Galois theory implies the following theorems.

Theorem 3.2.5 *An algebraic function y defined by Eq. (1) is representable by radicals over the field of rational functions if and only if its monodromy group is solvable.*

Theorem 3.2.6 *An algebraic function y defined by Eq. (1) is representable by k-radicals over the field of rational functions if and only if its monodromy group is k-solvable.*

V.I. Arnold proved (see [Alekseev 04]) by purely topological methods without using Galois theory that if an algebraic function is representable by radicals, then its monodromy group is solvable. Apart from the unsolvability of equations by radicals, Vladimir Igorevich proved the topological unsolvability of a whole series of classical mathematical problems. According to Arnold, a topological proof of unsolvability of a problem implies new and stronger corollaries.

I constructed a topological variant of Galois theory in which the monodromy group plays the role of the Galois group (see [Khovanskii 04b, Khovanskii 04a]). This topological variant of Galois theory is applicable to a wide class of multivalued functions of one complex variable. For instance, it is applicable to functions defined by differential Fuchsian equations and gives the strongest results about unsolvability of equations in finite terms. Connected ramified coverings over the Riemann sphere $\mathbb{C}P^1$ whose critical values lie in a fixed finite set O admit a complete description. Connected k-fold ramified coverings with marked points $\pi : (M, b) \to (\mathbb{C}P^1 \setminus O, a)$ are classified by subgroups of index k in the fundamental group $\pi_1(\mathbb{C}P^1 \setminus O)$. For every group G, we have the following lemma.

Lemma 3.2.7 *The classification of index-k subgroups of a group G is equivalent to the classification of transitive actions of the group G on a k-point set with a marked point.*

Proof Indeed, with a subgroup G_0 of index k in the group G one can associate the transitive action of the group G on the set of right cosets of the group G by the subgroup G_0. This set consists of k points, and the right coset of the identity element is a marked point. In the other direction, to every transitive action of the group G, one can assign the stabilizer G_0 of the marked point. This subgroup has index k in the group G.   □

The fundamental group $\pi_1(\mathbb{C}P^1 \setminus O, a)$ is a free group with a finite number of generators. It has finitely many distinct transitive actions on the set of k elements. All these actions can be described as follows.

Let us number the points of the set O. Suppose that this set contains $m + 1$ points. The fundamental group $\pi_1(\mathbb{C}P^1 \setminus O, a)$ is a free group generated by paths $\gamma_1, \ldots, \gamma_m$, where γ_i is a path going around the ith point of the set O. Take a set of k elements with one marked element. In the group $S(k)$ of permutations of this set choose m arbitrary elements $\sigma_1, \ldots, \sigma_m$. We are interested in ordered collections $\sigma_1, \ldots, \sigma_m$ satisfying a single relation: the group of permutations generated by these elements must be transitive. There is a finite number of collections $\sigma_1, \ldots, \sigma_m$. One can check each of them, and choose all collections generating transitive groups. With every such collection, one can associate a unique ramified covering $\pi : (M, b) \to (\mathbb{C}P^1, a)$ with a marked point. It corresponds to the stabilizer of the marked element under the homomorphism $F : \pi_1(\mathbb{C}P^1 \setminus O, a) \to S(k)$ that maps the generator γ_i to the element σ_i. Hence, in a finite number of steps, one can list all transitive actions $F : \pi_1(\mathbb{C}P^1 \setminus O, a) \to S(k)$ of the fundamental group $\pi_1(\mathbb{C}P^1 \setminus O, a)$ on the set of k elements.

Conjugations of the group $S(k)$ act on the finite set of homomorphisms $F : \pi_1(\mathbb{C}P^1 \setminus O, a) \to S(k)$ with transitive images. The orbits of a finite group action on a finite set can in principle be enumerated. Hence, conjugacy classes of the subgroups of index k in the fundamental group can also be listed in a finite number of steps.

Therefore, we obtain a complete geometric description of all possible Galois extensions of the field of rational functions in one variable. Note that in this description, we used the Riemann existence theorem. The Riemann existence theorem does not help to describe algebraic extensions of other fields, such as the field of rational numbers. The problem of describing algebraic extensions of the field of rational numbers is open. For instance, it is unknown in general whether there exists an extension of the field of rational numbers whose Galois group is a given finite group.

References

[Alekseev 04] V.B. Alekseev, Abel's Theorem in Problems and Solutions, in *Based on the Lectures of Professor V.I. Arnold* (Kluwer Academic, Norwell, 2004)

[Berger 87] M. Berger, *Geometry* (Springer, Berlin, 1987). Translated from the French by M. Cole and S. Levy. Universitext

[Khovanskii 04a] A.G. Khovanskii, Solvability of equations by explicit formulae. Appendix to [Alekseev 04]

[Khovanskii 04b] A.G. Khovanskii, On solvability and unsolvability of equations in explicit form. Russian Mathematical Surveys **59**(4), 661–736 (2004)

A. Khovanskii, *Galois Theory, Coverings, and Riemann Surfaces*,
DOI 10.1007/978-3-642-38841-5, © Springer-Verlag Berlin Heidelberg 2013

Index

A. Khovanskii, *Galois Theory, Coverings, and Riemann Surfaces*,
DOI 10.1007/978-3-642-38841-5, © Springer-Verlag Berlin Heidelberg 2013

Printed in the United States
By Bookmasters

Printed in the United States
By Bookmasters